Get better grades with . . .

Scientific Notebook™
for Windows® 95 and Windows NT® 4.0
ISBN: 0-534-34864-5. **$74.95**

Breakthrough interactive software for anyone who uses mathematics!

Whether you do simple arithmetic or solve complex partial differential equations, the solutions are just a mouse-click away. This one-of-a-kind scientific word-processor—with its special built-in version of the Maple® computer algebra system—can handle arithmetic, algebra, calculus, linear algebra, and more!

The only software tool that uses correct mathematical notation!

Scientific Notebook allows you to enter equations, create tables and matrices, import graphics, and graph in 2-D and 3-D within your documents. And it's easy to send and receive documents containing live mathematics on the World Wide Web. This combination gives you a unique tool for exploring, understanding, and explaining key mathematical and scientific concepts.

*The vast majority of my students are enthusiastic about **Scientific Notebook**. They find the difference between using **Scientific Notebook** and using a graphing calculator something like the difference between riding in a donkey cart and taking a ride in the space shuttle.*
—Johnathan Lewin, Kennesaw State University

To order a copy of *Scientific Notebook*, please contact your college store or place your order online at: http://www.scinotebook.com or fill out the order form and return with your payment.

ORDER FORM

_____Yes! Send me a copy of ***Scientific Notebook*™ for Windows®95 and Windows NT® 4.0**   (ISBN: 0-534-34864-5)

_____Copies x $74.95 = _____

Residents of: AL, AZ, CA, CT, CO, FL, GA, IL, IN, KS, KY, LA, MA, MD, MI, MN, MO, NC, NJ, NY, OH, PA, RI, SC, TN, TX, UT, VA, WA, WI must add appropriate state sales tax.

Subtotal _____
Tax _____
Handling $4.00
Total Due _____

Payment Options

_____ Check or money order enclosed

Bill my   ____VISA   ____MasterCard   ____American Express

Card Number: _____

Expiration Date: _____

Signature: _____

Please ship my order to: *(Credit card billing and shipping addresses must be the same)*

Name _____

Institution _____

Street Address_____

City _____   State _____   Zip+4_____

Telephone (   )_____   e-mail _____

Your credit card will not be billed until your order is shipped. Prices subject to change without notice. We will refund payment for unshipped out-of-stock titles after 120 days and for not-yet-published titles after 180 days unless an earlier date is requested in writing from you.

Mail to:

Brooks/Cole Publishing Company
Source Code 8BCTC054
511 Forest Lodge Road
Pacific Grove, California 93950-5098
Phone: (408) 373-0728; Fax: (408) 375-6414
e-mail: info@brookscole.com

Student Solutions Manual for Zill's
DIFFERENTIAL EQUATIONS
with Computer Lab Experiments

2nd Edition

W. Scott Wright
Loyola-Marymount University

Carol D. Wright

Brooks/Cole Publishing Company
I(T)P® *An International Thomson Publishing Company*

Pacific Grove • Albany • Belmont • Bonn • Boston • Cincinnati • Detroit
Johannesburg • London • Madrid • Melbourne • Mexico City • New York
Paris • Singapore • Tokyo • Toronto • Washington

 A GARY W. OSTEDT BOOK

Project Development Editor: *Elizabeth Rammel*
Editorial Associate: *Carol Benedict*
Marketing Manager: *Caroline Croley*
Production: *Dorothy Bell*

Cover Design: *Roy R. Neuhaus*
Cover Photo: *Telegraph Colour Library/FPG International*
Printing and Binding: *Patterson Printing*

COPYRIGHT © 1998 by Brooks/Cole Publishing Company
A division of International Thomson Publishing Inc.
I(T)P The ITP logo is a registered trademark under license.

For more information, contact:

BROOKS/COLE PUBLISHING COMPANY
511 Forest Lodge Road
Pacific Grove, CA 93950
USA

International Thomson Publishing Europe
Berkshire House 168-173
High Holborn
London WC1V 7AA
England

Thomas Nelson Australia
102 Dodds Street
South Melbourne, 3205
Victoria, Australia

Nelson Canada
1120 Birchmount Road
Scarborough, Ontario
Canada M1K 5G4

International Thomson Editores
Seneca 53
Col. Polanco
11560 México, D. F., México

International Thomson Publishing GmbH
Königswinterer Strasse 418
53227 Bonn
Germany

International Thomson Publishing Asia
221 Henderson Road
#05-10 Henderson Building
Singapore 0315

International Thomson Publishing Japan
Hirakawacho Kyowa Building, 3F
2-2-1 Hirakawacho
Chiyoda-ku, Tokyo 102
Japan

All rights reserved. No part of this work may be reproduced, stored in a retrieval system, or transcribed, in any form or by any means—electronic, mechanical, photocopying, recording, or otherwise—without the prior written permission of the publisher, Brooks/Cole Publishing Company, Pacific Grove, California 93950.

Printed in the United States of America
10 9 8 7 6 5 4 3 2 1
ISBN 0-534-35175-1

Table of Contents

1	Getting Started	1
2	First-Order Equations	3
3	Higher-Order Equations	15
4	Systems of First-Order Equations	41
5	The Laplace Transform	57
6	Series Solutions	73
Appendix I	Introduction to Matrices	88

1 Getting Started

Exercises 1.1

3. Nonlinear because of yy'

6. Nonlinear because of $\cos(t+y)$

9. Linear

12. From $y = \frac{6}{5} - \frac{6}{5}e^{-20t}$ we obtain $dy/dt = 24e^{-20t}$, so that

$$\frac{dy}{dt} + 20y = 24e^{-20t} + 20\left(\frac{6}{5} - \frac{6}{5}e^{-20t}\right) = 24.$$

15. First write the differential equation in the form $2xy + (x^2 + 2y)y' = 0$. Implicitly differentiating $x^2y + y^2 = 1$ we obtain $2xy + (x^2 + 2y)y' = 0$.

18. Differentiating $y = e^{-t^2}\int_0^t e^{u^2}\,du + c_1 e^{-t^2}$ we obtain

$$y' = e^{-t^2}e^{t^2} - 2te^{-t^2}\int_0^t e^{u^2}\,du - 2c_1 te^{-t^2} = 1 - 2te^{-t^2}\int_0^t e^{u^2}\,du - 2c_1 te^{-t^2}.$$

Substituting into the differential equation, we have

$$y' + 2ty = 1 - 2te^{-t^2}\int_0^t e^{u^2}\,du - 2c_1 te^{-t^2} + 2te^{-t^2}\int_0^t e^{u^2}\,du + 2c_1 te^{-t^2} = 1.$$

21. From $x = e^{-2t} + 3e^{6t}$ and $y = -e^{-2t} + 5e^{6t}$ we obtain

$$\frac{dx}{dt} = -2e^{-2t} + 18e^{6t} \quad \text{and} \quad \frac{dy}{dt} = 2e^{-2t} + 30e^{6t}.$$

Then

$$x + 3y = (e^{-2t} + 3e^{6t}) + 3(-e^{-2t} + 5e^{6t})$$
$$= -2e^{-2t} + 18e^{6t} = \frac{dx}{dt}$$

and

$$5x + 3y = 5(e^{-2t} + 3e^{6t}) + 3(-e^{-2t} + 5e^{6t})$$
$$= 2e^{-2t} + 30e^{6t} = \frac{dy}{dt}.$$

24. The function $y = \begin{cases} \sqrt{25-t^2}, & -5 < t < 0 \\ -\sqrt{25-t^2}, & 0 \le t < 5 \end{cases}$ is not continuous at $t = 0$ (the left hand limit is 5 and the right hand limit is -5,) and hence y' does not exist at $t = 0$.

1

Exercises 1.2

3. Using $x' = -c_1 \sin t + c_2 \cos t$ we obtain $c_1 = -1$ and $c_2 = 8$. The solution is $x = -\cos t + 8 \sin t$.

6. Using $x' = -c_1 \sin t + c_2 \cos t$ we obtain

$$\frac{\sqrt{2}}{2} c_1 + \frac{\sqrt{2}}{2} c_2 = \sqrt{2}$$

$$-\frac{\sqrt{2}}{2} c_1 + \frac{\sqrt{2}}{2} c_2 = 2\sqrt{2}.$$

Solving we find $c_1 = -1$ and $c_2 = 3$. The solution is $x = -\cos t + 3 \sin t$.

9. Using $y' = c_1 e^t - c_2 e^{-t}$ we obtain

$$e^{-1} c_1 + e c_2 = 5$$

$$e^{-1} c_1 - e c_2 = -5.$$

Solving we find $c_1 = 0$ and $c_2 = 5e^{-1}$. The solution is $y = 5e^{-(t+1)}$.

12. Two solutions are $y = 0$ and $y = t^2$.

15. For $f(t, y) = \dfrac{y}{t}$ we have $\dfrac{\partial f}{\partial y} = \dfrac{1}{t}$. Thus the differential equation will have a unique solution in any region where $t \neq 0$.

18. For $f(t, y) = \dfrac{t^2}{1 + y^3}$ we have $\dfrac{\partial f}{\partial y} = \dfrac{-3t^2 y^2}{(1 + y^3)^2}$. Thus the differential equation will have a unique solution in any region where $y \neq -1$.

For Problems 21 and 24 we identify $f(t, y) = \sqrt{y^2 - 9}$ and $\partial f / \partial y = y^2 / \sqrt{y^2 - 9}$. We further note that $f(t, y)$ is discontinuous for $|y| < 3$ and that $\partial f / \partial y$ is discontinuous for $|y| < 3$. We then apply Theorem 1.1.

21. The differential equation has a unique solution at $(1, 4)$.

24. The differential equation is not guaranteed to have a unique solution at $(-1, 1)$.

2 First-Order Equations

Exercises 2.1

3.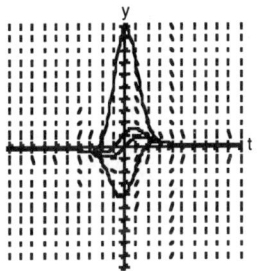

6.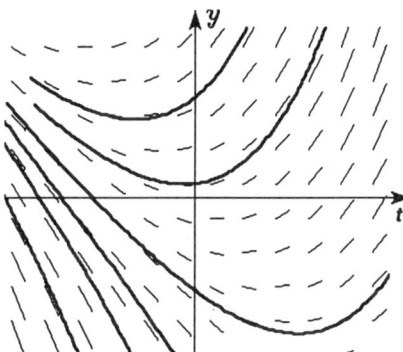

9.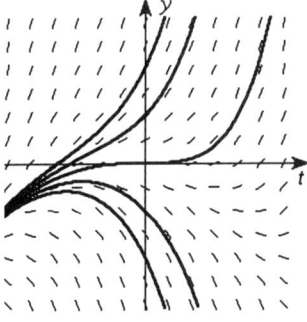

12.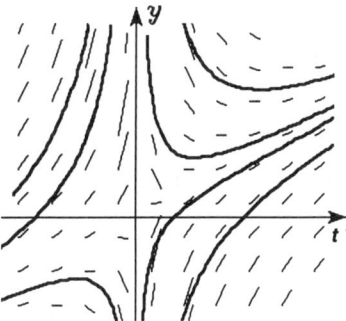

15. Solving $(x-2)^2 = 0$ we obtain the critical point 2.

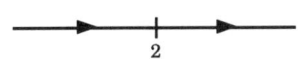

From the phase portrait we see that 2 is unstable.

18. Solving $x(2-x)(4-x) = 0$ we obtain the critical points 0, 2, and 4.

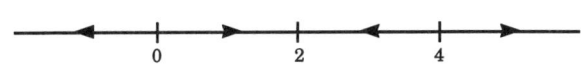

From the phase portrait we see that 2 is asymptotically stable, and 0 and 4 are unstable and repellers.

Exercises 2.1

21. Writing the differential equation in the form $dx/dt = x(1-x)(1+x)$ we see that critical points are located at $x = -1$, $x = 0$, and $x = 1$. The phase portrait is shown below.

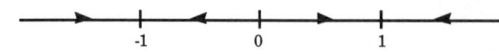

(a)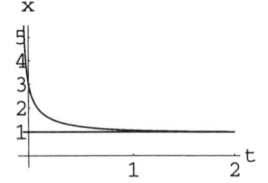

(b)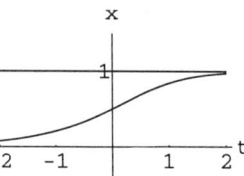

(c)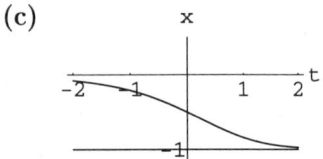

(d)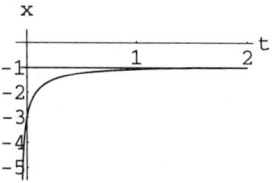

24. (a) From the phase portrait we see that critical points are α and β. Let $X(0) = X_0$.

If $X_0 < \alpha$, we see that $X \to \alpha$ as $t \to \infty$. If $\alpha < X_0 < \beta$, we see that $X \to \alpha$ as $t \to \infty$. If $X_0 > \beta$, we see that $X(t)$ increases in an unbounded manner, but more specific behavior of $X(t)$ as $t \to \infty$ is not known.

(b) When $\alpha = \beta$ the phase portrait is as shown.

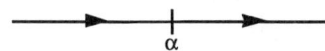

If $X_0 < \alpha$, then $X(t) \to \alpha$ as $t \to \infty$. If $X_0 > \alpha$, then $X(t)$ increases in an unbounded manner. This could happen in a finite amount of time. That is, the phase portrait does not indicate that X becomes unbounded as $t \to \infty$.

(c) When $k = 1$ and $\alpha = \beta$ the differential equation is $dX/dt = (\alpha - X)^2$. Separating variables and integrating we have

$$\frac{dX}{(\alpha - X)^2} = dt$$

$$\frac{1}{\alpha - X} = t + c$$

$$\alpha - X = \frac{1}{t + c}$$

$$X = \alpha - \frac{1}{t + c}.$$

Exercises 2.2

For $X(0) = \alpha/2$ we obtain
$$X(t) = \alpha - \frac{1}{t + 2/\alpha}.$$

For $X(0) = 2\alpha$ we obtain
$$X(t) = \alpha - \frac{1}{t - 1/\alpha}.$$

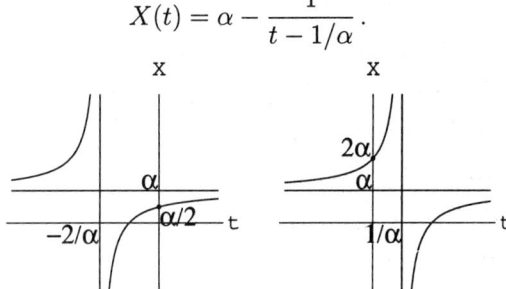

For $X_0 > \alpha$, $X(t)$ increases without bound up to $t = 1/\alpha$. For $t > 1/\alpha$, $X(t)$ increases but $X \to \alpha$ as $t \to \infty$.

Exercises 2.2

In many of the following problems we will encounter an expression of the form $\ln|g(y)| = f(t) + c$. To solve for $g(y)$ we exponentiate both sides of the equation. This yields $|g(y)| = e^{f(t)+c} = e^c e^{f(t)}$ which implies $g(y) = \pm e^c e^{f(t)}$. Letting $c_1 = \pm e^c$ we obtain $g(y) = c_1 e^{f(t)}$.

3. From $dy = -e^{-3t} dt$ we obtain $y = \frac{1}{3} e^{-3t} + c$.

6. From $\frac{1}{y} dy = -2x\, dx$ we obtain $\ln|y| = -x^2 + c$ or $y = c_1 e^{-x^2}$.

9. From $\left(y + 2 + \frac{1}{y}\right) dy = x^2 \ln x\, dx$ we obtain $\frac{y^2}{2} + 2y + \ln|y| = \frac{x^3}{3} \ln|x| - \frac{1}{9} x^3 + c$.

12. From $2y\, dy = -\frac{\sin 3t}{\cos^3 3t} dt = -\tan 3t \sec^2 3t\, dt$ we obtain $y^2 = -\frac{1}{6} \sec^2 3t + c$.

15. From $\frac{1}{S} dS = k\, dr$ we obtain $S = c e^{kr}$.

18. From $\frac{1}{N} dN = \left(t e^{t+2} - 1\right) dt$ we obtain $\ln|N| = t e^{t+2} - e^{t+2} - t + c$.

21. From $\frac{1}{x^2 + 1} dx = 4\, dt$ we obtain $\tan^{-1} x = 4t + c$. Using $x(\pi/4) = 1$ we find $c = -3\pi/4$. The solution of the initial-value problem is $\tan^{-1} x = 4t - \frac{3\pi}{4}$ or $x = \tan\left(4t - \frac{3\pi}{4}\right)$.

Exercises 2.2

24. From $\dfrac{1}{1-2y}\,dy = dt$ we obtain $-\dfrac{1}{2}\ln|1-2y| = t + c$ or $1 - 2y = c_1 e^{-2t}$. Using $y(0) = 5/2$ we find $c_1 = -4$. The solution of the initial-value problem is $1 - 2y = -4e^{-2t}$ or $y = 2e^{-2t} + \dfrac{1}{2}$.

27. Letting $y = ut$ we have
$$(t - ut)\,dt + t(u\,dt + t\,du) = 0$$
$$dt + t\,du = 0$$
$$\dfrac{dt}{t} + du = 0$$
$$\ln|t| + u = c$$
$$t\ln|t| + y = ct.$$

30. Letting $y = ut$ we have
$$\left(u^2 t^2 + ut^2\right)dt + t^2(u\,dt + t\,du) = 0$$
$$\left(u^2 + 2u\right)dt + t\,du = 0$$
$$\dfrac{dt}{t} + \dfrac{du}{u(u+2)} = 0$$
$$\ln|t| + \dfrac{1}{2}\ln|u| - \dfrac{1}{2}\ln|u+2| = c$$
$$\dfrac{t^2 u}{u+2} = c_1$$
$$t^2 \dfrac{y}{t} = c_1\left(\dfrac{y}{t} + 2\right)$$
$$t^2 y = c_1(y + 2t).$$

Exercises 2.3

3. For $y' + y = e^{3t}$ an integrating factor is $e^{\int dt} = e^t$ so that $\dfrac{d}{dt}\left[e^t y\right] = e^{4t}$ and $y = \dfrac{1}{4}e^{3t} + ce^{-t}$ for $-\infty < t < \infty$. The transient term is ce^{-t}.

6. For $y' + 2ty = t^3$ an integrating factor is $e^{\int 2t\,dt} = e^{t^2}$ so that $\dfrac{d}{dt}\left[e^{t^2} y\right] = t^3 e^{t^2}$ and $y = \dfrac{1}{2}t^2 - \dfrac{1}{2} + ce^{-t^2}$ for $-\infty < t < \infty$. The transient term is ce^{-t^2}.

Exercises 2.3

9. For $y' - \frac{1}{t}y = t \sin t$ an integrating factor is $e^{-\int (1/t)dt} = \frac{1}{t}$ so that $\frac{d}{dt}\left[\frac{1}{t}y\right] = \sin t$ and $y = ct - t\cos t$ for $0 < t < \infty$.

12. For $y' - \frac{t}{(1+t)}y = t$ an integrating factor is $e^{-\int [t/(1+t)]dt} = (t+1)e^{-t}$ so that $\frac{d}{dt}\left[(t+1)e^{-t}y\right] = t(t+1)e^{-t}$ and $y = -t - \frac{2t+3}{t+1} + \frac{ce^t}{t+1}$ for $-1 < t < \infty$.

15. For $\frac{dx}{dy} - \frac{4}{y}x = 4y^5$ an integrating factor is $e^{-\int (4/y)dy} = y^{-4}$ so that $\frac{d}{dy}\left[y^{-4}x\right] = 4y$ and $x = 2y^6 + cy^4$ for $0 < y < \infty$.

18. For $y' + (\cot t)y = \sec^2 t \csc t$ an integrating factor is $e^{\int \cot t\, dt} = \sin t$ so that $\frac{d}{dt}\left[(\sin t)y\right] = \sec^2 t$ and $y = \sec t + c\csc t$ for $0 < t < \pi/2$.

21. For $\frac{dr}{d\theta} + r\sec\theta = \cos\theta$ an integrating factor is $e^{\int \sec\theta\, d\theta} = \sec\theta + \tan\theta$ so that $\frac{d}{d\theta}[r(\sec\theta + \tan\theta)] = 1 + \sin\theta$ and $r(\sec\theta + \tan\theta) = \theta - \cos\theta + c$ for $-\pi/2 < \theta < \pi/2$.

24. For $\frac{dx}{dy} - \frac{1}{y}x = 2y$ an integrating factor is $e^{-\int (1/y)dy} = \frac{1}{y}$ so that $\frac{d}{dy}\left[\frac{1}{y}x\right] = 2$ and $x = 2y^2 + cy$ for $-\infty < y < \infty$. If $y(1) = 5$ then $c = -49/5$ and $x = 2y^2 - \frac{49}{5}y$.

27. For $x' + \frac{1}{t+1}x = \frac{\ln t}{t+1}$ an integrating factor is $e^{\int [1/(t+1)]dt} = t+1$ so that $\frac{d}{dt}[(t+1)x] = \ln t$ and $x = \frac{t}{t+1}\ln t - \frac{t}{t+1} + \frac{c}{t+1}$ for $0 < t < \infty$. If $x(1) = 10$ then $c = 21$ and $x = \frac{t}{t+1}\ln t - \frac{t}{t+1} + \frac{21}{t+1}$.

30. For $y' + y = f(t)$ an integrating factor is e^t so that

$$ye^t = \begin{cases} e^t + c_1, & 0 \le t \le 1; \\ -e^t + c_2, & t > 1. \end{cases}$$

If $y(0) = 1$ then $c_1 = 0$ and for continuity we must have $c_2 = 2e$ so that

$$y = \begin{cases} 1, & 0 \le t \le 1; \\ 2e^{1-t} - 1, & t > 1. \end{cases}$$

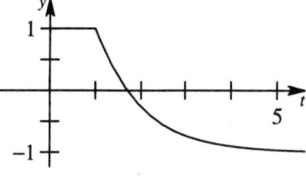

33. (a) An integrating factor for $y' - 2ty = -1$ is e^{-t^2}. Thus

$$\frac{d}{dt}[e^{-t^2}y] = -e^{-t^2}$$

$$e^{-t^2}y = -\int_0^t e^{-x^2}\, dx = -\frac{\sqrt{\pi}}{2}\operatorname{erf}(t) + c.$$

7

Exercises 2.3

From $y(0) = \sqrt{\pi}/2$, and noting that $\text{erf}(0) = 0$, we get $c = \sqrt{\pi}/2$. Thus

$$y = e^{t^2}\left(-\frac{\sqrt{\pi}}{2}\text{erf}(t) + \frac{\sqrt{\pi}}{2}\right) = \frac{\sqrt{\pi}}{2}e^{t^2}(1 - \text{erf}(t))$$

$$= \frac{\sqrt{\pi}}{2}e^{t^2}\text{erfc}(t).$$

(b) Using *Mathematica* we find $y(2) \approx 0.226339$.

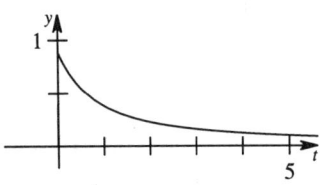

36. From $y' - y = e^t y^2$ and $w = y^{-1}$ we obtain $\dfrac{dw}{dt} + w = -e^t$. An integrating factor is e^t so that $e^t w = -\dfrac{1}{2}e^{2t} + c$ or $y^{-1} = -\dfrac{1}{2}e^t + ce^{-t}$.

Exercises 2.4

3. Let $P = P(t)$ be the population at time t. From $dP/dt = kT$ and $P(0) = P_0 = 500$ we obtain $P = 500e^{kt}$. Using $P(10) = 575$ we find $k = \frac{1}{10}\ln 1.15$. Then $P(30) = 500e^{3\ln 1.15} \approx 760$ years.

6. Let $N = N(t)$ be the amount at time t. From $dN/dt = kN$ and $N(0) = 100$ we obtain $N = 100e^{kt}$. Using $N(6) = 97$ we find $k = \frac{1}{6}\ln 0.97$. Then $N(24) = 100e^{(1/6)(\ln 0.97)24} = 100(0.97)^4 \approx 88.5$ mg.

9. Let $I = I(t)$ be the intensity, t the thickness, and $I(0) = I_0$. If $dI/dt = kI$ and $I(3) = .25I_0$ then $I = I_0 e^{kt}$, $k = \frac{1}{3}\ln .25$, and $I(15) = .00098I_0$.

12. Assume that $dT/dt = k(T - 5)$ so that $T = 5 + ce^{kt}$. If $T(1) = 55°$ and $T(5) = 30°$ then $k = -\frac{1}{4}\ln 2$ and $c = 59.4611$ so that $T(0) = 64.4611°$.

15. Assume $L\,di/dt + Ri = E(t)$, $L = .1$, $R = 50$, and $E(t) = 50$ so that $i = \frac{3}{5} + ce^{-500t}$. If $i(0) = 0$ then $c = -3/5$ and $\lim_{t\to\infty} i(t) = 3/5$.

18. Assume $R\,dq/dt + (1/c)q = E(t)$, $R = 1000$, $C = 5 \times 10^{-6}$, and $E(t) = 200$ so that $q = 1/1000 + ce^{-200t}$ and $i = -200ce^{-200t}$. If $i(0) = .4$ then $c = -1/500$, $q(.005) = .003$ coulombs, and $i(.005) = .1472$ amps. As $t \to \infty$ we have $q \to 1/1000$.

21. From $dA/dt = 4 - A/50$ we obtain $A = 200 + ce^{-t/50}$. If $A(0) = 30$ then $c = -170$ and $A = 200 - 170e^{-t/50}$.

24. From $\dfrac{dA}{dt} = 10 - \dfrac{10A}{500 - (10 - 5)t} = 10 - \dfrac{2A}{100 - t}$ we obtain $A = 1000 - 10t + c(100 - t)^2$. If $A(0) = 0$ then $c = -\dfrac{1}{10}$. The tank is empty in 100 minutes.

8

Exercises 2.4

27. (a) The differential equation is $\dfrac{dA}{dt} = k(M-A)$, where $k > 0$.

(b) The differential equation is $\dfrac{dA}{dt} = k_1(M-A) - k_2 A$, where $k_1 > 0$ and $k_2 > 0$.

(c) Separating variables and integrating the differential equation in part (a) we have

$$-\dfrac{dA}{M-A} = -k\,dt \quad \text{and} \quad \ln(M-A) = -kt + c,$$

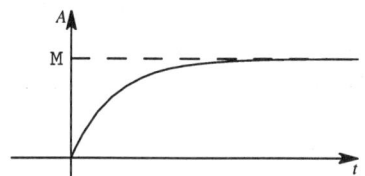

so $A = M - c_1 e^{-kt}$. Using $A(0) = 0$ we find $c_1 = M$, so $A = M(1 - e^{-kt})$. As $t \to \infty$ we see that $A \to M$. This means that over a long period of time practically all of the material will be memorized.

Write the differential equation in part (b) in the form

$$dA/dt + (k_1 + k_2)A = k_1 M.$$

Then an integrating factor is $e^{(k_1+k_2)t}$, and

$$\dfrac{d}{dt}\left[e^{(k_1+k_2)t} A\right] = k_1 M e^{(k_1+k_2)t}$$

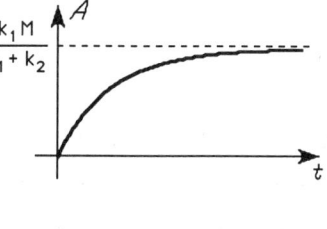

$$e^{(k_1+k_2)t} A = \dfrac{k_1 M}{k_1 + k_2} e^{(k_1+k_2)t} + c$$

$$A = \dfrac{k_1 M}{k_1 + k_2} + c e^{-(k_1+k_2)t}.$$

Using $A(0) = 0$ we find $c = -\dfrac{k_1 M}{k_1 + k_2}$ and $A = \dfrac{k_1 M}{k_1 + k_2}\left(1 - e^{-(k_1+k_2)t}\right)$. As $t \to \infty$ we see that $A \to \dfrac{k_1 M}{k_1 + k_2}$. If $k_1 > 0$ and $k_2 > 0$, the material will never be completely memorized.

30. From $\dfrac{dN}{d} = N(a - bN)$ and $N(0) = 500$ we obtain $N = \dfrac{500a}{500b + (a - 500b)e^{-at}}$. Since $\lim\limits_{t\to\infty} N = \dfrac{a}{b} = 50{,}000$ and $N(1) = 1000$ we have $a = .7033$, $b = .00014$, and $N = \dfrac{50{,}000}{1 + 99e^{-.7033t}}$.

33. If $\alpha \ne \beta$, $\dfrac{dX}{dt} = k(\alpha - X)(\beta - X)$, and $X(0) = 0$ then $\left(\dfrac{1/(\beta-\alpha)}{\alpha - X} + \dfrac{1/(\alpha-\beta)}{\beta - X}\right) dX = k\,dt$ so that $X = \dfrac{\alpha - c\beta e^{(\alpha-\beta)kt}}{1 - c e^{(\alpha-\beta)kt}}$. If $\alpha = \beta$ then $\dfrac{1}{(\alpha - X)^2} dX = k\,dt$ and $X = \alpha - \dfrac{1}{kt + c}$.

36. (a) Let ρ be the weight density of the water and V the volume of the object. Archimedes' principle states that the upward buoyant force has magnitude equal to the weight of the water displaced.

9

Exercises 2.4

Taking the positive direction to be down, the differential equation is

$$m\frac{dv}{dt} = mg - kv^2 - \rho V.$$

(b) Using separation of variables we have

$$\frac{m\,dv}{(mg - \rho V) - kv^2} = dt$$

$$\frac{m}{\sqrt{k}}\frac{\sqrt{k}\,dv}{(\sqrt{mg - \rho V})^2 - (\sqrt{k}\,v)^2} = dt$$

$$\frac{m}{\sqrt{k}}\frac{1}{\sqrt{mg - \rho V}}\tanh^{-1}\frac{\sqrt{k}\,v}{\sqrt{mg - \rho V}} = t + c.$$

Thus

$$v(t) = \sqrt{\frac{mg - \rho V}{k}}\tanh\left(\frac{\sqrt{kmg - k\rho V}}{m}t + c_1\right).$$

(c) Since $\tanh t \to 1$ as $t \to \infty$, the terminal velocity is $\sqrt{(mg - \rho V)/k}$.

Exercises 2.5

3. (a) Euler's method with h = 0.1

t_n	y_n
1.00	5.0000
1.10	3.8000
1.20	2.9800
1.30	2.4260
1.40	2.0582
1.50	1.8207

(b) Euler's method with h = 0.05

t_n	y_n
1.00	5.0000
1.05	4.4000
1.10	3.8950
1.15	3.4708
1.20	3.1151
1.25	2.8179
1.30	2.5702
1.35	2.3647
1.40	2.1950
1.45	2.0557
1.50	1.9424

Exercises 2.5

6. (a) Euler's method with h = 0.1

t_n	y_n
0.00	1.0000
0.10	1.1000
0.20	1.2220
0.30	1.3753
0.40	1.5735
0.50	1.8371

(b) Euler's method method with h = 0.05

t_n	y_n
0.00	1.0000
0.05	1.0500
0.10	1.1053
0.15	1.1668
0.20	1.2360
0.25	1.3144
0.30	1.4039
0.35	1.5070
0.40	1.6267
0.45	1.7670
0.50	1.9332

9. (a) Euler's method with h = 0.1

t_n	y_n
0.00	0.5000
0.10	0.5250
0.20	0.5431
0.30	0.5548
0.40	0.5613
0.50	0.5639

(b) Euler's method with h = 0.05

t_n	y_n
0.00	0.5000
0.05	0.5125
0.10	0.5232
0.15	0.5322
0.20	0.5395
0.25	0.5452
0.30	0.5496
0.35	0.5527
0.40	0.5547
0.45	0.5559
0.50	0.5565

12. (a) Euler's method with h = 0.1

t_n	y_n
1.00	0.5000
1.10	0.5250
1.20	0.5499
1.30	0.5747
1.40	0.5991
1.50	0.6231

(b) Euler's method method with h = 0.05

t_n	y_n
1.00	0.5000
1.05	0.5125
1.10	0.5250
1.15	0.5375
1.20	0.5499
1.25	0.5623
1.30	0.5746
1.35	0.5868
1.40	0.5989
1.45	0.6109
1.50	0.6228

Exercises 2.5

15. (a)

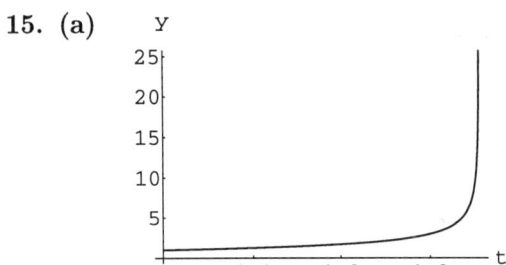

(b) Comparison of Numerical Methods with h = 0.1

t_n	Euler	Improved Euler
1.00	1.0000	1.0000
1.10	1.2000	1.2469
1.20	1.4938	1.6668
1.30	1.9711	2.6427
1.40	2.9060	8.7988

18. (a) Using the improved Euler method we obtain $y(0.1) \approx y_1 = 1.22$.

(b) Using $y''' = 8e^{2t}$ we see that the local truncation error is

$$y'''(c)\frac{h^3}{6} = 8e^{2c}\frac{(0.1)^3}{6} = 0.001333e^{2c}.$$

Since e^{2t} is an increasing function, $e^{2c} \leq e^{2(0.1)} = e^{0.2}$ for $0 \leq c \leq 0.1$. Thus an upper bound for the local truncation error is $0.001333e^{0.2} = 0.001628$.

(c) Since $y(0.1) = e^{0.2} = 1.221403$, the actual error is $y(0.1) - y_1 = 0.001403$ which is less than 0.001628.

(d) Using the improved Euler method with $h = 0.05$ we obtain $y(0.1) \approx y_2 = 1.221025$.

(e) The error in part (d) is $1.221403 - 1.221025 = 0.000378$. With global truncation error $O(h^2)$, when the step size is halved we expect the error for $h = 0.05$ to be one-fourth the error for $h = 0.1$. Comparing 0.000378 with 0.001403 we see that this is the case.

21. (a) Using $y'' = 38e^{-3(t-1)}$ we see that the local truncation error is

$$y''(c)\frac{h^2}{2} = 38e^{-3(c-1)}\frac{h^2}{2} = 19h^2 e^{-3(c-1)}.$$

(b) Since $e^{-3(t-1)}$ is a decreasing function for $1 \leq t \leq 1.5$, $e^{-3(c-1)} \leq e^{-3(1-1)} = 1$ for $1 \leq c \leq 1.5$ and

$$y''(c)\frac{h^2}{2} \leq 19(0.1)^2(1) = 0.19.$$

(c) Using the Euler method with $h = 0.1$ we obtain $y(1.5) \approx 1.8207$. With $h = 0.05$ we obtain $y(1.5) \approx 1.9424$.

(d) Since $y(1.5) = 2.0532$, the error for $h = 0.1$ is $E_{0.1} = 0.2325$, while the error for $h = 0.05$ is $E_{0.05} = 0.1109$. With global truncation error $O(h)$ we expect $E_{0.1}/E_{0.05} \approx 2$. We actually have $E_{0.1}/E_{0.05} = 2.10$.

Exercises 2.5

24. **(a)** Using $y''' = \dfrac{2}{(t+1)^3}$ we see that the local truncation error is

$$y'''(c)\frac{h^3}{6} = \frac{1}{(c+1)^3}\frac{h^3}{3}.$$

(b) Since $\dfrac{1}{(t+1)^3}$ is a decreasing function for $0 \le t \le 0.5$, $\dfrac{1}{(c+1)^3} \le \dfrac{1}{(0+1)^3} = 1$ for $0 \le c \le 0.5$ and

$$y'''(c)\frac{h^3}{6} \le (1)\frac{(0.1)^3}{3} = 0.000333.$$

(c) Using the improved Euler method with $h = 0.1$ we obtain $y(0.5) \approx 0.405281$. With $h = 0.05$ we obtain $y(0.5) \approx 0.405419$.

(d) Since $y(0.5) = 0.405465$, the error for $h = 0.1$ is $E_{0.1} = 0.000184$, while the error for $h = 0.05$ is $E_{0.05} = 0.000046$. With global truncation error $O(h^2)$ we expect $E_{0.1}/E_{0.05} \approx 4$. We actually have $E_{0.1}/E_{0.05} = 3.98$.

27. Runge-Kutta method with $h = 0.1$

t_n	y_n
1.00	5.0000
1.10	3.9724
1.20	3.2284
1.30	2.6945
1.40	2.3163
1.50	2.0533

30. Runge-Kutta method with $h = 0.1$

t_n	y_n
0.00	1.0000
0.10	1.1115
0.20	1.2530
0.30	1.4397
0.40	1.6961
0.50	2.0670

33. Runge-Kutta method with $h = 0.1$

t_n	y_n
0.00	0.5000
0.10	0.5213
0.20	0.5358
0.30	0.5443
0.40	0.5482
0.50	0.5493

36. Runge-Kutta method with $h = 0.1$

t_n	y_n
0.00	0.5000
0.10	0.5250
0.20	0.5498
0.30	0.5744
0.40	0.5987
0.50	0.6225

Exercises 2.5

39. (a) Runge-Kutta Method

t_n	h = 0.05	h = 0.1
1.00	1.0000	1.0000
1.05	1.1112	
1.10	1.2511	1.2511
1.15	1.4348	
1.20	1.6934	1.6934
1.25	2.1047	
1.30	2.9560	2.9425
1.35	7.8981	
1.40	1.0608E+15	903.0282

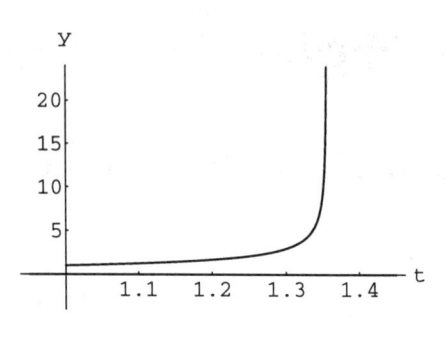

(b) [graph of y vs t with curve rising sharply near t = 1.4, axes labeled with y values 5, 10, 15, 20 and t values 1.1, 1.2, 1.3, 1.4]

42. (a) Using $y^{(5)} = -1026e^{-3(t-1)}$ we see that the local truncation error is

$$\left| y^{(5)}(c) \frac{h^5}{120} \right| = 8.55 h^5 e^{-3(c-1)}.$$

(b) Since $e^{-3(t-1)}$ is a decreasing function for $1 \leq t \leq 1.5$, $e^{-3(c-1)} \leq e^{-3(1-1)} = 1$ for $1 \leq c \leq 1.5$ and

$$y^{(5)}(c) \frac{h^5}{120} \leq 8.55(0.1)^5(1) = 0.0000855.$$

(c) Using the fourth-order Runge-Kutta method with $h = 0.1$ we obtain $y(1.5) \approx 2.053338827$. With $h = 0.05$ we obtain $y(1.5) \approx 2.053222989$.

3 Higher-Order Equations

Exercises 3.1

3. From $y = c_1 e^{4t} + c_2 e^{-t}$ we find $y' = 4c_1 e^{4t} - c_2 e^{-t}$. Then $y(0) = c_1 + c_2 = 1$, $y'(0) = 4c_1 - c_2 = 2$ so that $c_1 = 3/5$ and $c_2 = 2/5$. The solution is $y = \frac{3}{5}e^{4t} + \frac{2}{5}e^{-t}$.

6. From $y = c_1 + c_2 t^2$ we find $y' = 2c_2 t$. Then $y(0) = c_1 = 0$, $y'(0) = 2c_2 \cdot 0 = 0$ and $y'(0) = 1$ is not possible. Since $a_2(t) = t$ is 0 at $t = 0$, Theorem 3.1 is not violated.

9. From $y = c_1 e^t \cos t + c_2 e^t \sin t$ we find $y' = c_1 e^t (-\sin t + \cos t) + c_2 e^t (\cos t + \sin t)$.

  (a) We have $y(0) = c_1 = 1$, $y'(0) = c_1 + c_2 = 0$ so that $c_1 = 1$ and $c_2 = -1$. The solution is $y = e^t \cos t - e^t \sin t$.

  (b) We have $y(0) = c_1 = 1$, $y(\pi) = -c_1 e^\pi = -1$, which is not possible.

  (c) We have $y(0) = c_1 = 1$, $y(\pi/2) = c_2 e^{\pi/2} = 1$ so that $c_1 = 1$ and $c_2 = e^{-\pi/2}$. The solution is $y = e^t \cos t + e^{-\pi/2} e^t \sin t$.

  (d) We have $y(0) = c_1 = 0$, $y(\pi) = -c_1 e^\pi = 0$ so that $c_1 = 0$ and c_2 is arbitrary. Solutions are $y = c_2 e^t \sin t$, for any real numbers c_2.

12. Since $a_0(t) = \tan t$ and $t_0 = 0$ the problem has a unique solution for $-\pi/2 < t < \pi/2$.

15. Since $(-1/5)5 + (1)\cos^2 t + (1)\sin^2 t = 0$ the functions are linearly dependent.

18. From the graphs of $f_1(t) = 2 + t$ and $f_2(t) = 2 + |t|$ we see that the functions are linearly independent since they cannot be multiples of each other.

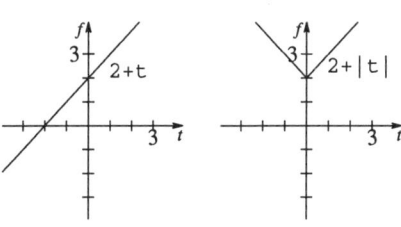

21. The functions satisfy the differential equation and are linearly independent since
$$W\left(e^{-3t}, e^{4t}\right) = 7e^t \neq 0$$
for $-\infty < t < \infty$. The general solution is
$$y = c_1 e^{-3t} + c_2 e^{4t}.$$

24. The functions satisfy the differential equation and are linearly independent since
$$W\left(e^{t/2}, te^{t/2}\right) = e^t \neq 0$$

Exercises 3.1

for $-\infty < t < \infty$. The general solution is
$$y = c_1 e^{t/2} + c_2 t e^{t/2}.$$

27. The functions satisfy the differential equation and are linearly independent since
$$W\left(t, t^{-2}, t^{-2}\ln t\right) = 9t^{-6} \neq 0$$
for $0 < t < \infty$. The general solution is
$$y = c_1 t + c_2 t^{-2} + c_3 t^{-2}\ln t.$$

30. The functions $y_1 = \cos t$ and $y_2 = \sin t$ form a fundamental set of solutions of the homogeneous equation, and $y_p = t\sin t + (\cos t)\ln(\cos t)$ is a particular solution of the nonhomogeneous equation.

33. By the superposition principle for nonhomogeneous equations a particular solution of $y'' - 6y' + 5y = 5t^2 + 3t - 16 - 9e^{2t}$ is $y_p = t^2 + 3t + 3e^{2t}$. A particular solution of the second equation is
$$y_p = -2y_{p_2} \cdot \frac{1}{9}y_{p_1} = -2t^2 - 6t - \frac{1}{3}e^{2t}.$$

Exercises 3.2

3. From $m^2 + 9 = 0$ we obtain $m = 3i$ and $m = -3i$ so that $y = c_1\cos 3t + c_2\sin 3t$.

6. From $m^2 - 3m + 2 = 0$ we obtain $m = 1$ and $m = 2$ so that $y = c_1 e^t + c_2 e^{2t}$.

9. From $m^2 + 3m - 5 = 0$ we obtain $m = -3/2 \pm \sqrt{29}/2$ so that $y = c_1 e^{(-3+\sqrt{29})t/2} + c_2 e^{(-3-\sqrt{29})t/2}$.

12. From $8m^2 + 2m - 1 = 0$ we obtain $m = 1/4$ and $m = -1/2$ so that $y = c_1 e^{t/4} + c_2 e^{-t/2}$.

15. From $m^3 - 4m^2 - 5m = 0$ we obtain $m = 0$, $m = 5$, and $m = -1$ so that
$$y = c_1 + c_2 e^{5t} + c_3 e^{-t}.$$

18. From $m^3 + 5m^2 = 0$ we obtain $m = 0$, $m = 0$, and $m = -5$ so that
$$y = c_1 + c_2 t + c_3 e^{-5t}.$$

21. From $m^3 + m^2 - 2 = 0$ we obtain $m = 1$ and $m = -1 \pm i$ so that
$$y = c_1 e^t + e^{-t}(c_2 \cos t + c_3 \sin t).$$

24. From $m^3 - 6m^2 + 12m - 8 = 0$ we obtain $m = 2$, $m = 2$, and $m = 2$ so that
$$y = c_1 e^{2t} + c_2 t e^{2t} + c_3 t^2 e^{2t}.$$

16

Exercises 3.3

27. From $16m^4 + 24m^2 + 9 = 0$ we obtain $m = \pm\sqrt{3}\,i/2$ and $m = \pm\sqrt{3}\,i/2$ so that
$$y = c_1 \cos \sqrt{3}\,x/2 + c_2 \sin \sqrt{3}\,x/2 + c_3 x \cos \sqrt{3}\,x/2 + c_4 x \sin \sqrt{3}\,x/2.$$

30. From $m^5 - 2m^4 + 17m^3 = 0$ we obtain $m = 0$, $m = 0$, $m = 0$, and $m = 1 \pm 4i$ so that
$$y = c_1 + c_2 x + c_3 x^2 + e^x (c_4 \cos 4x + c_5 \sin 4x).$$

33. From $2m^2 - 2m + 1 = 0$ we obtain $m = 1/2 \pm i/2$ so that $y = e^{t/2}(c_1 \cos t/2 + c_2 \sin t/2)$. If $y(0) = -1$ and $y'(0) = 0$ then $c_1 = -1$, $\frac{1}{2}c_1 + \frac{1}{2}c_2 = 0$, so $c_1 = -1$, $c_2 = 1$, and $y = e^{t/2}\left(\sin \frac{1}{2}t - \cos \frac{1}{2}t\right)$.

36. From $4m^2 - 4m - 3 = 0$ we obtain $m = -1/2$ and $m = 3/2$ so that $y = c_1 e^{-t/2} + c_2 e^{3t/2}$. If $y(0) = 1$ and $y'(0) = 5$ then $c_1 + c_2 = 1$, $-\frac{1}{2}c_1 + \frac{3}{2}c_2 = 5$, so $c_1 = -7/4$, $c_2 = 11/4$, and $y = -\frac{7}{4}e^{-t/2} + \frac{11}{4}e^{3t/2}$.

39. From $m^3 + 12m^2 + 36m = 0$ we obtain $m = 0$, $m = -6$, and $m = -6$ so that $y = c_1 + c_2 e^{-6t} + c_3 t e^{-6t}$. If $y(0) = 0$, $y'(0) = 1$, and $y''(0) = -7$ then
$$c_1 + c_2 = 0, \quad -6c_2 + c_3 = 1, \quad 36c_2 - 12c_3 = -7,$$
so $c_1 = 5/36$, $c_2 = -5/36$, $c_3 = 1/6$, and $y = \frac{5}{36} - \frac{5}{36}e^{-6t} + \frac{1}{6}te^{-6t}$.

42. From $m^4 - 1 = 0$ we obtain $m = 1$, $m = -1$, and $m = \pm i$ so that $y = c_1 e^t + c_2 e^{-t} + c_3 \cos t + c_4 \sin t$. If $y(0) = 0$, $y'(0) = 0$, $y''(0) = 0$, and $y'''(0) = 1$ then
$$c_1 + c_2 + c_3 = 0, \quad c_1 - c_2 + c_4 = 0, \quad c_1 + c_2 - c_3 = 0, \quad c_1 - c_2 - c_4 = 1,$$
so $c_1 = 1/4$, $c_2 = -1/4$, $c_3 = 0$, $c_4 = -1/2$, and
$$y = \frac{1}{4}e^t - \frac{1}{4}e^{-t} - \frac{1}{2}\sin t.$$

45. From $m^2 + 1 = 0$ we obtain $m = \pm i$ so that $y = c_1 \cos t + c_2 \sin t$. If $y'(0) = 0$ and $y'(\pi/2) = 2$ then $c_1 = -2$, $c_2 = 0$ and $y = -2\cos t$.

Exercises 3.3

3. For $x = e^t$ we have $y = x' = e^t = x$. Similarly, for $x = 2e^t$ we have $y = x' = 2e^t = x$. Thus, the particular solutions $x = e^t$ and $x = 2e^t$ both correspond to the trajectory $y = x$ for $x > 0$.

6. Letting $y = dx/dt$ we see that the corresponding system is
$$\frac{dx}{dt} = y$$
$$\frac{dy}{dt} = -10x + 2y.$$

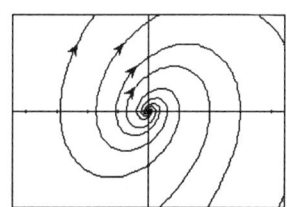

9. The roots of the auxiliary equation are $m_1 = 2$ and $m_2 = 5$. $(0,0)$ is an unstable node.

Exercises 3.3

12. The roots of the auxiliary equation are $m_1 = m_2 = 3$. $(0,0)$ is an unstable node.

15. (a)

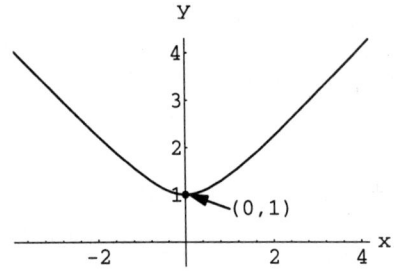

(b)

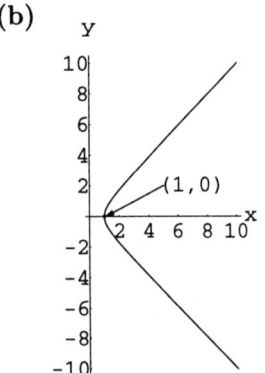

(c)

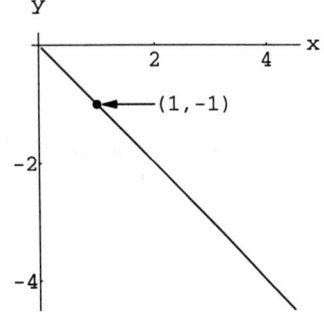

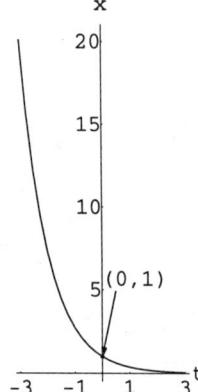

//
Exercises 3.4

3. From $m^2 - 10m + 25 = 0$ we find $m_1 = m_2 = 5$. Then $y_c = c_1 e^{5t} + c_2 t e^{5t}$ and we assume $y_p = At + B$. Substituting into the differential equation we obtain $25A = 30$ and $-10A + 25B = 3$. Then $A = \frac{6}{5}$, $B = \frac{3}{5}$, $y_p = \frac{6}{5}t + \frac{3}{5}$, and
$$y = c_1 e^{5t} + c_2 t e^{5t} + \frac{6}{5}t + \frac{3}{5}.$$

6. From $m^2 - 8m + 20 = 0$ we find $m_1 = 2 + 4i$ and $m_2 = 2 - 4i$. Then $y_c = e^{2t}(c_1 \cos 4t + c_2 \sin 4t)$ and we assume $y_p = At^2 + Bt + C + (Dt + E)e^t$. Substituting into the differential equation we obtain

$$2A - 8B + 20C = 0$$

$$-6D + 13E = 0$$

$$-16A + 20B = 0$$

$$13D = -26$$

$$20A = 100.$$

Then $A = 5$, $B = 4$, $C = \frac{11}{10}$, $D = -2$, $E = -\frac{12}{13}$, $y_p = 5t^2 + 4t + \frac{11}{10} + \left(-2t - \frac{12}{13}\right)e^t$ and

$$y = e^{2t}(c_1 \cos 4t + c_2 \sin 4t) + 5t^2 + 4t + \frac{11}{10} + \left(-2t - \frac{12}{13}\right)e^t.$$

9. From $m^2 - m = 0$ we find $m_1 = 1$ and $m_2 = 0$. Then $y_c = c_1 e^t + c_2$ and we assume $y_p = At$. Substituting into the differential equation we obtain $-A = -3$. Then $A = 3$, $y_p = 3t$ and $y = c_1 e^t + c_2 + 3t$.

12. From $m^2 - 16 = 0$ we find $m_1 = 4$ and $m_2 = -4$. Then $y_c = c_1 e^{4t} + c_2 e^{-4t}$ and we assume $y_p = At e^{4t}$. Substituting into the differential equation we obtain $8A = 2$. Then $A = \frac{1}{4}$, $y_p = \frac{1}{4} t e^{4t}$ and
$$y = c_1 e^{4t} + c_2 e^{-4t} + \frac{1}{4} t e^{4t}.$$

15. From $m^2 + 1 = 0$ we find $m_1 = i$ and $m_2 = -i$. Then $y_c = c_1 \cos t + c_2 \sin t$ and we assume $y_p = (At^2 + Bt)\cos t + (Ct^2 + Dt)\sin t$. Substituting into the differential equation we obtain $4C = 0$, $2A + 2D = 0$, $-4A = 2$, and $-2B + 2C = 0$. Then $A = -\frac{1}{2}$, $B = 0$, $C = 0$, $D = \frac{1}{2}$, $y_p = -\frac{1}{2}t^2 \cos t + \frac{1}{2}t \sin t$, and
$$y = c_1 \cos t + c_2 \sin t - \frac{1}{2} t^2 \cos t + \frac{1}{2} t \sin t.$$

Exercises 3.4

18. From $m^2 - 2m + 2 = 0$ we find $m_1 = 1 + i$ and $m_2 = 1 - i$. Then $y_c = e^t(c_1 \cos t + c_2 \sin t)$ and we assume $y_p = Ae^{2t} \cos t + Be^{2t} \sin t$. Substituting into the differential equation we obtain $A + 2B = 1$ and $-2A + B = -3$. Then $A = \frac{7}{5}$, $B = -\frac{1}{5}$, $y_p = \frac{7}{5}e^{2t}\cos t - \frac{1}{5}e^{2t}\sin t$ and

$$y = e^t(c_1 \cos t + c_2 \sin t) + \frac{7}{5}e^{2t}\cos t - \frac{1}{5}e^{2t}\sin t.$$

21. From $m^3 - 6m^2 = 0$ we find $m_1 = m_2 = 0$ and $m_3 = 6$. Then $y_c = c_1 + c_2 t + c_3 e^{6t}$ and we assume $y_p = At^2 + B\cos t + C\sin t$. Substituting into the differential equation we obtain $-12A = 3$, $6B - C = -1$, and $B + 6C = 0$. Then $A = -\frac{1}{4}$, $B = -\frac{6}{37}$, $C = \frac{1}{37}$, $y_p = -\frac{1}{4}t^2 - \frac{6}{37}\cos t + \frac{1}{37}\sin t$, and

$$y = c_1 + c_2 t + c_3 e^{6t} - \frac{1}{4}t^2 - \frac{6}{37}\cos t + \frac{1}{37}\sin t.$$

24. From $m^3 - m^2 - 4m + 4 = 0$ we find $m_1 = 1$, $m_2 = 2$, and $m_3 = -2$. Then $y_c = c_1 e^t + c_2 e^{2t} + c_3 e^{-2t}$ and we assume $y_p = A + Bte^t + Cte^{2t}$. Substituting into the differential equation we obtain $4A = 5$, $-3B = -1$, and $4C = 1$. Then $A = \frac{5}{4}$, $B = \frac{1}{3}$, $C = \frac{1}{4}$, $y_p = \frac{5}{4} + \frac{1}{3}te^t + \frac{1}{4}te^{2t}$, and

$$y = c_1 e^t + c_2 e^{2t} + c_3 e^{-2t} + \frac{5}{4} + \frac{1}{3}te^t + \frac{1}{4}te^{2t}.$$

27. We have $y_c = c_1 e^{-t/5} + c_2$ and we assume $y_p = At^2 + Bt$. Substituting into the differential equation we find $A = -3$ and $B = 30$. Thus $y = c_1 e^{-t/5} + c_2 - 3t^2 + 30t$. From the initial conditions we obtain $c_1 = 200$ and $c_2 = -200$, so

$$y = 200^{-t/5} - 200 - 3t^2 + 30t.$$

30. The auxiliary equation is $m^2 + 1 = 0$, so $y_c = c_1 \cos t + c_2 \sin t$ and

$$W = \begin{vmatrix} \cos t & \sin t \\ -\sin t & \cos t \end{vmatrix} = 1.$$

Identifying $f(t) = \tan t$ we obtain

$$u_1' = -\sin t \tan t = \frac{\cos^2 t - 1}{\cos t} = \cos t - \sec t$$

$$u_2' = \sin t.$$

Then $u_1 = \sin t - \ln|\sec t + \tan t|$, $u_2 = -\cos t$, and

$$y = c_1 \cos t + c_2 \sin t + \cos t (\sin t - \ln|\sec t + \tan t|) - \cos t \sin t$$

for $-\pi/2 < t < \pi/2$.

33. The auxiliary equation is $m^2 + 1 = 0$, so $y_c = c_1 \cos t + c_2 \sin t$ and

$$W = \begin{vmatrix} \cos t & \sin t \\ -\sin t & \cos t \end{vmatrix} = 1.$$

Exercises 3.4

Identifying $f(t) = \cos^2 t$ we obtain

$$u_1' = -\sin t \cos^2 t$$
$$u_2' = \cos^3 t = \cos t \left(1 - \sin^2 t\right).$$

Then $u_1 = \frac{1}{3}\cos^3 t$, $u_2 = \sin t - \frac{1}{3}\sin^3 t$, and

$$y = c_1 \cos t + c_2 \sin t + \frac{1}{3}\cos^4 t + \sin^2 t - \frac{1}{3}\sin^4 t$$

$$= c_1 \cos t + c_2 \sin t + \frac{1}{3}\left(\cos^2 t + \sin^2 t\right)\left(\cos^2 t - \sin^2 t\right) + \sin^2 t$$

$$= c_1 \cos t + c_2 \sin t + \frac{1}{3}\cos^2 t + \frac{2}{3}\sin^2 t$$

$$= c_1 \cos t + c_2 \sin t + \frac{1}{3} + \frac{1}{3}\sin^2 t$$

for $-\infty < t < \infty$.

36. The auxiliary equation is $m^2 - 1 = 0$, so $y_c = c_1 e^t + c_2 e^{-t}$ and

$$W = \begin{vmatrix} e^t & e^{-t} \\ e^t & -e^{-t} \end{vmatrix} = -2.$$

Identifying $f(t) = \sinh 2t$ we obtain

$$u_1' = -\frac{1}{4}e^{-3t} + \frac{1}{4}e^t$$

$$u_2' = \frac{1}{4}e^{-t} - \frac{1}{4}e^{3t}.$$

Then

$$u_1 = \frac{1}{12}e^{-3t} + \frac{1}{4}e^t$$

$$u_2 = -\frac{1}{4}e^{-t} - \frac{1}{12}e^{3t}.$$

and

$$y = c_1 e^t + c_2 e^{-t} + \frac{1}{12}e^{-2t} + \frac{1}{4}e^{2t} - \frac{1}{4}e^{-2t} - \frac{1}{12}e^{2t}$$

$$= c_1 e^t + c_2 e^{-t} + \frac{1}{6}\left(e^{2t} - e^{-2t}\right)$$

$$= c_1 e^t + c_2 e^{-t} + \frac{1}{3}\sinh 2t$$

for $-\infty < t < \infty$.

21

Exercises 3.4

39. The auxiliary equation is $m^2 + 3m + 2 = (m+1)(m+2) = 0$, so $y_c = c_1 e^{-t} + c_2 e^{-2t}$ and
$$W = \begin{vmatrix} e^{-t} & e^{-2t} \\ -e^{-t} & -2e^{-2t} \end{vmatrix} = -e^{-3t}.$$

Identifying $f(t) = 1/(1 + e^t)$ we obtain
$$u_1' = \frac{e^t}{1 + e^t}$$
$$u_2' = -\frac{e^{2t}}{1 + e^t} = \frac{e^t}{1 + e^t} - e^t.$$

Then $u_1 = \ln(1 + e^t)$, $u_2 = \ln(1 + e^t) - e^t$, and
$$y = c_1 e^{-t} + c_2 e^{-2t} + e^{-t}\ln(1 + e^t) + e^{-2t}\ln(1 + e^t) - e^{-t}$$
$$= c_3 e^{-t} + c_2 e^{-2t} + (1 + e^{-t})e^{-t}\ln(1 + e^t)$$

for $-\infty < t < \infty$.

42. The auxiliary equation is $m^2 - 2m + 1 = (m-1)^2 = 0$, so $y_c = c_1 e^t + c_2 t e^t$ and
$$W = \begin{vmatrix} e^t & te^t \\ e^t & te^t + e^t \end{vmatrix} = e^{2t}.$$

Identifying $f(t) = e^t \tan^{-1} t$ we obtain
$$u_1' = -\frac{te^t e^t \tan^{-1} t}{e^{2t}} = -t\tan^{-1} t$$
$$u_2' = \frac{e^t e^t \tan^{-1} t}{e^{2t}} = \tan^{-1} t.$$

Then
$$u_1 = -\frac{1 + t^2}{2}\tan^{-1} t + \frac{t}{2}$$
$$u_2 = t\tan^{-1} t - \frac{1}{2}\ln(1 + t^2)$$

and
$$y = c_1 e^t + c_2 t e^t + \left(-\frac{1 + t^2}{2}\tan^{-1} t + \frac{t}{2}\right) e^t + \left(t\tan^{-1} t - \frac{1}{2}\ln(1 + t^2)\right) t e^t$$
$$= c_1 e^t + c_3 t e^t + \frac{1}{2} e^t \left[(t^2 - 1)\tan^{-1} t - \ln(1 + t^2)\right]$$

for $-\infty < t < \infty$.

45. The auxiliary equation is $m^2 + 2m + 1 = (m+1)^2 = 0$, so $y_c = c_1 e^{-t} + c_2 t e^{-t}$ and
$$W = \begin{vmatrix} e^{-t} & te^{-t} \\ -e^{-t} & -te^{-t} + e^{-t} \end{vmatrix} = e^{-2t}.$$

Identifying $f(t) = e^{-t} \ln t$ we obtain

$$u_1' = -\frac{te^{-t}e^{-t}\ln t}{e^{-2t}} = -t\ln t$$

$$u_2' = \frac{e^{-t}e^{-t}\ln t}{e^{-2t}} = \ln t.$$

Then

$$u_1 = -\frac{1}{2}t^2 \ln t + \frac{1}{4}t^2$$

$$u_2 = t\ln t - t$$

and

$$y = c_1 e^{-t} + c_2 t e^{-t} - \frac{1}{2}t^2 e^{-t} \ln t + \frac{1}{4}t^2 e^{-t} + t^2 e^{-t} \ln t - t^2 e^{-t}$$

$$= c_1 e^{-t} + c_2 t e^{-t} + \frac{1}{2}t^2 e^{-t} \ln t - \frac{3}{4}t^2 e^{-t}$$

for $t > 0$.

48. The auxiliary equation is $4m^2 - 4m + 1 = (2m-1)^2 = 0$, so $y_c = c_1 e^{t/2} + c_2 t e^{t/2}$ and

$$W = \begin{vmatrix} e^{t/2} & te^{t/2} \\ \frac{1}{2}e^{t/2} & \frac{1}{2}te^{t/2} + e^{t/2} \end{vmatrix} = e^t.$$

Identifying $f(t) = \frac{1}{4}e^{t/2}\sqrt{1-t^2}$ we obtain

$$u_1' = -\frac{te^{t/2}e^{t/2}\sqrt{1-t^2}}{4e^t} = -\frac{1}{4}t\sqrt{1-t^2}$$

$$u_2' = \frac{e^{t/2}e^{t/2}\sqrt{1-t^2}}{4e^t} = \frac{1}{4}\sqrt{1-t^2}.$$

Then

$$u_1 = \frac{1}{12}\left(1-t^2\right)^{3/2}$$

$$u_2 = \frac{t}{8}\sqrt{1-t^2} + \frac{1}{8}\sin^{-1}t$$

and

$$y = c_1 e^{t/2} + c_2 t e^{t/2} + \frac{1}{12}e^{t/2}\left(1-t^2\right)^{3/2} + \frac{1}{8}t^2 e^{t/2}\sqrt{1-t^2} + \frac{1}{8}te^{t/2}\sin^{-1}t$$

for $-1 \leq t \leq 1$.

51. The auxiliary equation is $m^3 - 2m^2 - m + 2 = (m-1)(m-2)(m+1) = 0$, so $y_c = c_1 e^t + c_2 e^{2t} + c_3 e^{-t}$ and

$$W = \begin{vmatrix} e^t & e^{2t} & e^{-t} \\ e^t & 2e^{2t} & -e^{-t} \\ e^t & 4e^{2t} & e^{-t} \end{vmatrix} = 6e^{2t}.$$

Exercises 3.4

Identifying $f(t) = e^{3t}$ we obtain

$$u_1' = \frac{1}{6e^{2t}} W_1 = \frac{1}{6e^{2t}} \begin{vmatrix} 0 & e^{2t} & e^{-t} \\ 0 & 2e^{2t} & -e^{-t} \\ e^{3t} & 4e^{2t} & e^{-t} \end{vmatrix} = \frac{-3e^{4t}}{6e^{2t}} = -\frac{1}{2}e^{2t}$$

$$u_2' = \frac{1}{6e^{2t}} W_2 = \frac{1}{6e^{2t}} \begin{vmatrix} e^t & 0 & e^{-t} \\ e^t & 0 & -e^{-t} \\ e^t & e^{3t} & e^{-t} \end{vmatrix} = \frac{2e^{3t}}{6e^{2t}} = \frac{1}{3}e^t$$

$$u_3' = \frac{1}{6e^{2t}} W_1^3 = \frac{1}{6e^{2t}} \begin{vmatrix} e^t & e^{2t} & 0 \\ e^t & 2e^{2t} & 0 \\ e^t & 4e^{2t} & e^{3t} \end{vmatrix} = \frac{e^{6t}}{6e^{2t}} = \frac{1}{6}e^{4t}.$$

Then $u_1 = -\frac{1}{4}e^{2t}$, $u_2 = \frac{1}{3}e^t$, and $u_3 = \frac{1}{24}e^{4t}$, and

$$y = c_1 e^t + c_2 e^{2t} + c_3 e^{-t} - \frac{1}{4}e^{3t} + \frac{1}{3}e^{3t} + \frac{1}{24}e^{3t}$$

$$= c_1 e^t + c_2 e^{2t} + c_3 e^{-t} + \frac{1}{8}e^{3t}$$

for $-\infty < t < \infty$.

54. The auxiliary equation is $2m^2 + m - 1 = (2m - 1)(m + 1) = 0$, so $y_c = c_1 e^{t/2} + c_2 e^{-t}$ and

$$W = \begin{vmatrix} e^{t/2} & e^{-t} \\ \frac{1}{2}e^{t/2} & -e^{-t} \end{vmatrix} = -\frac{3}{2}e^{-t/2}.$$

Identifying $f(t) = (t+1)/2$ we obtain

$$u_1' = \frac{1}{3}e^{-t/2}(t+1)$$

$$u_2' = -\frac{1}{3}e^t(t+1).$$

Then

$$u_1 = -e^{-t/2}\left(\frac{2}{3}t - 2\right)$$

$$u_2 = -\frac{1}{3}te^t.$$

Thus

$$y = c_1 e^{t/2} + c_2 e^{-t} - t - 2$$

and

$$y' = \frac{1}{2}c_1 e^{t/2} - c_2 e^{-t} - 1.$$

The initial conditions imply
$$c_1 - c_2 - 2 = 1$$
$$\frac{1}{2}c_1 - c_2 - 1 = 0.$$

Thus $c_1 = 8/3$ and $c_2 = 1/3$, and
$$y = \frac{8}{3}e^{t/2} + \frac{1}{3}e^{-t} - t - 2.$$

Exercises 3.5

3. The auxiliary equation is $m^2 = 0$ so that $y = c_1 + c_2 \ln t$.

6. The auxiliary equation is $m^2 + 4m + 3 = (m+1)(m+3) = 0$ so that $y = c_1 t^{-1} + c_2 t^{-3}$.

9. The auxiliary equation is $25m^2 + 1 = 0$ so that $y = c_1 \cos\left(\frac{1}{5}\ln t\right) + c_2 \sin\left(\frac{1}{5}\ln t\right)$.

12. The auxiliary equation is $m^2 + 7m + 6 = (m+1)(m+6) = 0$ so that $y = c_1 t^{-1} + c_2 t^{-6}$.

15. The auxiliary equation is $3m^2 + 3m + 1 = 0$ so that $y = t^{-1/2}\left[c_1 \cos\left(\frac{\sqrt{3}}{6}\ln t\right) + c_2 \sin\left(\frac{\sqrt{3}}{6}\ln t\right)\right]$.

18. Assuming that $y = t^m$ and substituting into the differential equation we obtain
$$m(m-1)(m-2) + m - 1 = m^3 - 3m^2 + 3m - 1 = (m-1)^3 = 0.$$
Thus
$$y = c_1 t + c_2 t \ln t + c_3 t (\ln t)^2.$$

21. Assuming that $y = t^m$ and substituting into the differential equation we obtain
$$m(m-1)(m-2)(m-3) + 6m(m-1)(m-2) = m^4 - 7m^2 + 6m = m(m-1)(m-2)(m+3) = 0.$$
Thus
$$y = c_1 + c_2 t + c_3 t^2 + c_4 t^{-3}.$$

24. The auxiliary equation is $m^2 - 6m + 8 = (m-2)(m-4) = 0$, so that
$$y = c_1 t^2 + c_2 t^4 \quad \text{and} \quad y' = 2c_1 t + 4c_2 t^3.$$
The initial conditions imply
$$4c_1 + 16c_2 = 32$$
$$4c_1 + 32c_2 = 0.$$
Thus, $c_1 = 16$, $c_2 = -2$, and $y = 16t^2 - 2t^4$.

Exercises 3.5

27. We use the substitution $x = -t$ since the initial conditions are on the interval $(-\infty, 0)$. In this case

$$\frac{dy}{dx} = \frac{dy}{dt}\frac{dt}{dx} = -\frac{dy}{dt}$$

and

$$\frac{d^2y}{dx^2} = \frac{d}{dx}\left(\frac{dy}{dx}\right) = \frac{d}{dx}\left(-\frac{dy}{dt}\right) = -\frac{d}{dx}(y') = -\frac{dy'}{dt}\frac{dt}{dx} = -\frac{d^2y}{dt^2}\frac{dt}{dx} = \frac{d^2y}{dt^2}.$$

The differential equation and initial conditions become

$$4x^2\frac{d^2y}{dx^2} + y = 0; \quad y(x)\Big|_{x=1} = 2, \quad y'(x)\Big|_{x=1} = -4.$$

The auxiliary equation is $4m^2 - 4m + 1 = (2m-1)^2 = 0$, so that

$$y = c_1 x^{1/2} + c_2 x^{1/2}\ln x \quad \text{and} \quad y' = \frac{1}{2}c_1 x^{-1/2} + c_2\left(x^{-1/2} + \frac{1}{2}x^{-1/2}\ln x\right).$$

The initial conditions imply $c_1 = 2$ and $1 + c_2 = -4$. Thus

$$y = 2x^{1/2} - 5x^{1/2}\ln x = 2(-t)^{1/2} - 5(-t)^{1/2}\ln(-t), \quad t < 0.$$

30. The auxiliary equation is $m^2 - 5m = m(m-5) = 0$ so that $y_c = c_1 + c_2 t^5$ and

$$W(1, t^5) = \begin{vmatrix} 1 & t^5 \\ 0 & 5t^4 \end{vmatrix} = 5t^4.$$

Identifying $f(t) = t^3$ we obtain $u_1' = -\frac{1}{5}t^4$ and $u_2' = 1/5t$. Then $u_1 = -\frac{1}{25}t^5$, $u_2 = \frac{1}{5}\ln t$, and

$$y = c_1 + c_2 t^5 - \frac{1}{25}t^5 + \frac{1}{5}t^5 \ln t = c_1 + c_3 t^5 + \frac{1}{5}t^5 \ln t.$$

33. The auxiliary equation is $m^2 - 2m + 1 = (m-1)^2 = 0$ so that $y_c = c_1 t + c_2 t \ln t$ and

$$W(t, t\ln t) = \begin{vmatrix} t & t\ln t \\ 1 & 1 + \ln t \end{vmatrix} = t.$$

Identifying $f(t) = 2/t$ we obtain $u_1' = -2\ln t/t$ and $u_2' = 2/t$. Then $u_1 = -(\ln t)^2$, $u_2 = 2\ln t$, and

$$y = c_1 t + c_2 t \ln t - t(\ln t)^2 + 2t(\ln t)^2$$
$$= c_1 t + c_2 t \ln t + t(\ln t)^2.$$

Exercises 3.6

In Problems 36 and 39 we use the following results: When $t = e^x$ or $x = \ln t$, then

$$\frac{dy}{dt} = \frac{1}{t}\frac{dy}{dx} \quad \text{and} \quad \frac{d^2y}{dt^2} = \frac{1}{t^2}\left[\frac{d^2y}{dx^2} - \frac{dy}{dx}\right].$$

36. Substituting into the differential equation we obtain

$$\frac{d^2y}{dx^2} - 5\frac{dy}{dx} + 6y = 2x.$$

The auxiliary equation is $m^2 - 5m + 6 = (m-2)(m-3) = 0$ so that $y_c = c_1 e^{2x} + c_2 e^{3x}$. Using undetermined coefficients we try $y_p = Ax + B$. This leads to $(-5A + 6B) + 6Ax = 2x$, so that $A = 1/3$, $B = 5/18$, and

$$y = c_1 e^{2x} + c_2 e^{3x} + \frac{1}{3}x + \frac{5}{18} = c_1 t^2 + c_2 t^3 + \frac{1}{3}\ln t + \frac{5}{18}.$$

39. Substituting into the differential equation we obtain

$$\frac{d^2y}{dx^2} + 8\frac{dy}{dx} - 20y = 5e^{-3x}.$$

The auxiliary equation is $m^2 + 8m - 20 = (m+10)(m-2) = 0$ so that $y_c = c_1 e^{-10x} + c_2 e^{2x}$. Using undetermined coefficients we try $y_p = Ae^{-3x}$. This leads to $-35Ae^{-3x} = 5e^{-3x}$, so that $A = -1/7$ and

$$y = c_1 e^{-10x} + c_2 e^{2x} - \frac{1}{7}e^{-3x} = c_1 t^{-10} + c_2 t^2 - \frac{1}{7}t^{-3}.$$

Exercises 3.6

3. From $\frac{3}{4}x'' + 72x = 0$, $x(0) = -1/4$, and $x'(0) = 0$ we obtain $x = -\frac{1}{4}\cos 4\sqrt{6}\,t$.

6. From $50x'' + 200x = 0$, $x(0) = 0$, and $x'(0) = -10$ we obtain $x = -5\sin 2t$ and $x' = -10\cos 2t$.

9. From $\frac{1}{4}x'' + x = 0$, $x(0) = 1/2$, and $x'(0) = 3/2$ we obtain

$$x = \frac{1}{2}\cos 2t + \frac{3}{4}\sin 2t = \frac{\sqrt{13}}{4}\sin(2t + 0.588).$$

12. From $x' + 9x = 0$, $x(0) = -1$, and $x'(0) = -\sqrt{3}$ we obtain

$$x = -\cos 3t - \frac{\sqrt{3}}{3}\sin 3t = \frac{2}{\sqrt{3}}\sin\left(37 + \frac{4\pi}{3}\right)$$

and $x' = 2\sqrt{3}\cos(3t + 4\pi/3)$. If $x' = 3$ then $t = -7\pi/18 + 2n\pi/3$ and $t = -\pi/2 + 2n\pi/3$ for $n = 1$, 2, 3, ...

Exercises 3.6

15. From $\frac{1}{8}x'' + x' + 2x = 0$, $x(0) = -1$, and $x'(0) = 8$ we obtain $x = 4te^{-4t} - e^{-4t}$ and $x' = 8e^{-4t} - 16te^{-4t}$. If $x = 0$ then $t = 1/4$ second. If $x' = 0$ then $t = 1/2$ second and the extreme displacement is $x = e^{-2}$ feet.

18. (a) $x = \frac{1}{3}e^{-8t}\left(4e^{6t} - 1\right)$ is never zero; the extreme displacement is $x(0) = 1$ meter.

(b) $x = \frac{1}{3}e^{-8t}\left(5 - 2e^{6t}\right) = 0$ when $t = \frac{1}{6}\ln\frac{5}{2} \approx 0.153$ second; if $x' = \frac{4}{3}e^{-8t}\left(e^{6t} - 10\right) = 0$ then $t = \frac{1}{6}\ln 10 \approx 0.384$ second and the extreme displacement is $x = -0.232$ meter.

21. From $\frac{5}{16}x'' + \beta x' + 5x = 0$ we find that the roots of the auxiliary equation are $m = -\frac{8}{5}\beta \pm \frac{4}{5}\sqrt{4\beta^2 - 25}$.

(a) If $4\beta^2 - 25 > 0$ then $\beta > 5/2$.

(b) If $4\beta^2 - 25 = 0$ then $\beta = 5/2$.

(c) If $4\beta^2 - 25 < 0$ then $0 < \beta < 5/2$.

24. (a) If $x'' + 2x' + 5x = 12\cos 2t + 3\sin 2t$, $x(0) = -1$, and $x'(0) = 5$ then $x_c = e^{-t}(c_1\cos 2t + c_2\sin 2t)$ and $x_p = 3\sin 2t$ so that the equation of motion is $x = e^{-t}\cos 2t + 3\sin 2t$.

(b) (c)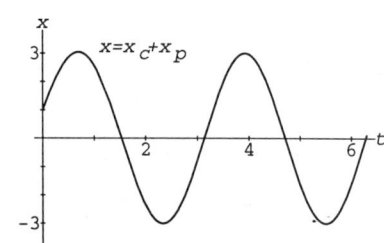

27. From $2x'' + 32x = 68e^{-2t}\cos 4t$, $x(0) = 0$, and $x'(0) = 0$ we obtain $x_c = c_1\cos 4t + c_2\sin 4t$ and $x_p = \frac{1}{2}e^{-2t}\cos 4t - 2e^{-2t}\sin 4t$ so that

$$x = -\frac{1}{2}\cos 4t + \frac{9}{4}\sin 4t + \frac{1}{2}e^{-2t}\cos 4t - 2e^{-2t}\sin 4t.$$

30. (a) From $100x'' + 1600x = 1600\sin 8t$, $x(0) = 0$, and $x'(0) = 0$ we obtain $x_c = c_1\cos 4t + c_2\sin 4t$ and $x_p = -\frac{1}{3}\sin 8t$ so that

$$x = \frac{2}{3}\sin 4t - \frac{1}{3}\sin 8t.$$

(b) If $x = \frac{1}{3}\sin 4t(2 - 2\cos 4t) = 0$ then $t = n\pi/4$ for $n = 0, 1, 2, \ldots$.

(c) If $x' = \frac{8}{3}\cos 4t - \frac{8}{3}\cos 8t = \frac{8}{3}(1-\cos 4t)(1+2\cos 4t) = 0$ then $t = \pi/3 + n\pi/2$ and $t = \pi/6 + n\pi/2$ for $n = 0, 1, 2, \ldots$ at the extreme values. *Note:* There are many other values of t for which $x' = 0$.

(d) $x(\pi/6 + n\pi/2) = \sqrt{3}/2$ cm. and $x(\pi/3 + n\pi/2) = -\sqrt{3}/2$ cm.

28

(e)

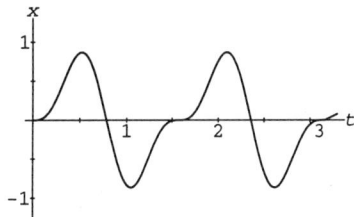

33. Solving $\frac{5}{3}q'' + 10q' + 30q = 300$ we obtain $q(t) = e^{-3t}(c_1 \cos 3t + c_2 \sin 3t) + 10$. The initial conditions $q(0) = q'(0) = 0$ imply $c_1 = c_2 = -10$. Thus

$$q(t) = 10 - 10e^{-3t}(\cos 3t + \sin 3t) \quad \text{and} \quad i(t) = 60e^{-3t} \sin 3t.$$

Solving $i(t) = 0$ we see that the maximum charge occurs when $t = \pi/3$ and $q(\pi/3) \approx 10.432$ coulombs.

36. (a) The steady-state solution $q_p(t)$ is a particular solution of the differential equation

$$L\frac{d^2q}{dt^2} + R\frac{dq}{dt} + \frac{1}{C}q = E_0 \sin \gamma t.$$

Using the method of undetermined coefficients, we assume a particular solution of the form $q_p(t) = A \sin \gamma t + B \cos \gamma t$. Substituting this expression into the differential equation, simplifying, and equating coefficients give

$$A = \frac{E_0(L\gamma - 1/C\gamma)}{-\gamma[L^2\gamma^2 - (2L/C + 1/C^2\gamma^2) + R^2]}, \qquad B = \frac{E_0 R}{-\gamma[L^2\gamma^2 - (2L/C + 1/C^2\gamma^2) + R^2]}.$$

It is convenient to express A and B in terms of some new symbols. If

$$X = L\gamma - \frac{1}{C\gamma} \quad \text{then} \quad X^2 = L^2\gamma^2 - \frac{2L}{C} + \frac{1}{C^2\gamma^2}.$$

If

$$Z = \sqrt{X^2 + R^2} \quad \text{then} \quad Z^2 = L^2\gamma^2 - \frac{2L}{C} + \frac{1}{C^2\gamma^2} + R^2.$$

Therefore

$$A = \frac{E_0 X}{-\gamma Z^2} \quad \text{and} \quad B = \frac{E_0 R}{-\gamma Z^2},$$

so the steady-state charge is

$$q_p(t) = -\frac{E_0 X}{\gamma Z^2} \sin \gamma t - \frac{E_0 R}{\gamma Z^2} \cos \gamma t.$$

(b) From

$$i_p(t) = \frac{E_0}{Z}\left(\frac{R}{Z} \sin \gamma t - \frac{X}{Z} \cos \gamma t\right)$$

Exercises 3.6

and $Z = \sqrt{X^2 + R^2}$ we see that the amplitude of $i_p(t)$ is

$$A = \sqrt{\frac{E_0^2 R^2}{Z^4} + \frac{E_0^2 X^2}{Z^4}} = \frac{E_0}{Z^2}\sqrt{R^2 + X^2} = \frac{E_0}{Z}.$$

Exercises 3.7

3. (a) The general solution is

$$y(x) = c_1 + c_2 x + c_3 x^2 + c_4 x^3 + \frac{w_0}{24EI} x^4.$$

The boundary conditions are $y(0) = 0$, $y'(0) = 0$, $y(L) = 0$, $y''(L) = 0$. The first two conditions give $c_1 = 0$ and $c_2 = 0$. The conditions at $x = L$ give the system

$$c_3 L^2 + c_4 L^3 + \frac{w_0}{24EI} L^4 = 0$$

$$2c_3 + 6c_4 L + \frac{w_0}{2EI} L^2 = 0.$$

Solving, we obtain $c_3 = w_0 L^2/16EI$ and $c_4 = -5w_0 L/48EI$. The deflection is

$$y(x) = \frac{w_0}{48EI}(3L^2 x^2 - 5Lx^3 + 2x^4).$$

(b)

6. (a) $y_{max} = y(L/2) = \dfrac{5w_0 L^4}{384EI}$

(b) The maximum deflection of the beam in Example 1 is $y(L/2) = (w_0/24EI)L^4/16 = w_0 L^4/384EI$, which is 1/5 of the maximum displacement of the beam in Problem 2.

9. For $\lambda \leq 0$ the only solution of the boundary-value problem is $y = 0$. For $\lambda > 0$ we have

$$y = c_1 \cos\sqrt{\lambda}\, x + c_2 \sin\sqrt{\lambda}\, x.$$

Now $y(0) = 0$ implies $c_1 = 0$, so

$$y(\pi) = c_2 \sin\sqrt{\lambda}\, \pi = 0$$

Exercises 3.7

gives
$$\sqrt{\lambda}\,\pi = n\pi \quad \text{or} \quad \lambda = n^2, \ n = 1, 2, 3, \ldots.$$

The eigenvalues n^2 correspond to the eigenfunctions $y = \sin nx$ for $n = 1, 2, 3, \ldots$.

12. For $\lambda \leq 0$ the only solution of the boundary-value problem is $y = 0$. For $\lambda > 0$ we have
$$y = c_1 \cos\sqrt{\lambda}\,x + c_2 \sin\sqrt{\lambda}\,x.$$

Now $y(0) = 0$ implies $c_1 = 0$, so
$$y'\left(\frac{\pi}{2}\right) = c_2\sqrt{\lambda}\cos\sqrt{\lambda}\,\frac{\pi}{2} = 0$$

gives
$$\sqrt{\lambda}\,\frac{\pi}{2} = \frac{(2n-1)\pi}{2} \quad \text{or} \quad \lambda = (2n-1)^2, \ n = 1, 2, 3, \ldots.$$

The eigenvalues $(2n-1)^2$ correspond to the eigenfunctions $y = \sin(2n-1)x$.

15. The auxiliary equation has solutions
$$m = \frac{1}{2}\left(-2 \pm \sqrt{4-4(\lambda+1)}\right) = -1 \pm \sqrt{-\lambda}.$$

For $\lambda < 0$ we have
$$y = e^{-x}\left(c_1 \cosh\sqrt{-\lambda}\,x + c_2 \sinh\sqrt{-\lambda}\,x\right).$$

The boundary conditions imply
$$y(0) = c_1 = 0$$
$$y(5) = c_2 e^{-5}\sinh 5\sqrt{-\lambda} = 0$$

so $c_1 = c_2 = 0$ and the only solution of the boundary-value problem is $y = 0$.

For $\lambda = 0$ we have
$$y = c_1 e^{-x} + c_2 x e^{-x}$$

and the only solution of the boundary-value problem is $y = 0$.

For $\lambda > 0$ we have
$$y = e^{-x}\left(c_1 \cos\sqrt{\lambda}\,x + c_2 \sin\sqrt{\lambda}\,x\right).$$

Now $y(0) = 0$ implies $c_1 = 0$, so
$$y(5) = c_2 e^{-5}\sin 5\sqrt{\lambda} = 0$$

gives
$$5\sqrt{\lambda} = n\pi \quad \text{or} \quad \lambda = \frac{n^2\pi^2}{25}, \ n = 1, 2, 3, \ldots.$$

The eigenvalues $n^2\pi^2/25$ correspond to the eigenfunctions $y = e^{-x}\sin\frac{n\pi}{5}x$ for $n = 1, 2, 3, \ldots$.

Exercises 3.7

18. For $\lambda = 0$ the only solution of the boundary-value problem is $y = 0$. For $\lambda \neq 0$ we have
$$y = c_1 \cos \lambda x + c_2 \sin \lambda x.$$
Now $y(0) = 0$ implies $c_1 = 0$, so
$$y'(3\pi) = c_2 \lambda \cos 3\pi\lambda = 0$$
gives
$$3\pi\lambda = \frac{(2n-1)\pi}{2} \quad \text{or} \quad \lambda = \frac{2n-1}{6}, \quad n = 1, 2, 3, \ldots.$$
The eigenvalues $(2n-1)/6$ correspond to the eigenfunctions $y = \sin \frac{2n-1}{6} x$ for $n = 1, 2, 3, \ldots$.

21. For $\lambda = 0$ the general solution is $y = c_1 + c_2 \ln x$. Now $y' = c_2/x$, so $y'(1) = c_2 = 0$ and $y = c_1$. Since $y'(e^2) = 0$ for any c_1 we see that $y(x) = 1$ is an eigenfunction corresponding to the eigenvalue $\lambda = 0$.
For $\lambda < 0$, $y = c_1 x^{-\sqrt{-\lambda}} + c_2 x^{\sqrt{-\lambda}}$. The initial conditions imply $c_1 = c_2 = 0$, so $y(x) = 0$.
For $\lambda > 0$, $y = c_1 \cos(\sqrt{\lambda} \ln x) + c_2 \sin(\sqrt{\lambda} \ln x)$. Now
$$y' = -c_1 \frac{\sqrt{\lambda}}{x} \sin(\sqrt{\lambda} \ln x) + c_2 \frac{\sqrt{\lambda}}{x} \cos(\sqrt{\lambda} \ln x),$$
and $y'(1) = c_2\sqrt{\lambda} = 0$ implies $c_2 = 0$. Finally, $y'(e^2) = -(c_1\sqrt{\lambda}/e^2)\sin(2\sqrt{\lambda}) = 0$ implies $\lambda = n^2\pi^2/4$ for $n = 1, 2, 3, \ldots$. The corresponding eigenfunctions are
$$y = \cos\left(\frac{n\pi}{2} \ln x\right).$$

24. (a) The general solution of the differential equation is
$$y = c_1 \cos \sqrt{\frac{P}{EI}}\, x + c_2 \sin \sqrt{\frac{P}{EI}}\, x + \delta.$$
Since the column is embedded at $x = 0$, the initial conditions are $y(0) = y'(0) = 0$. If $\delta = 0$ this implies that $c_1 = c_2 = 0$ and $y(x) = 0$. That is, there is no deflection.

(b) If $\delta \neq 0$, the initial conditions give, in turn, $c_1 = -\delta$ and $c_2 = 0$. Then
$$y = \delta\left(1 - \cos\sqrt{\frac{P}{EI}}\, x\right).$$
In order to satisfy the condition $y(L) = \delta$ we must have
$$\delta = \delta\left(1 - \cos\sqrt{\frac{P}{EI}}\, L\right) \quad \text{or} \quad \cos\sqrt{\frac{P}{EI}}\, L = 0.$$

This gives $\sqrt{P/EI}\, L = n\pi/2$ for $n = 1, 2, 3, \ldots$. The smallest value of P_n, the Euler load, is then

$$\sqrt{\frac{P_1}{EI}}\, L = \frac{\pi}{2} \quad \text{or} \quad P_1 = \frac{1}{4}\left(\frac{\pi^2 EI}{L^2}\right).$$

27. The auxiliary equation is $m^2 + m = m(m+1) = 0$ so that $u(r) = c_1 r^{-1} + c_2$. The boundary conditions $u(a) = u_0$ and $u(b) = u_1$ yield the system $c_1 a^{-1} + c_2 = u_0$, $c_1 b^{-1} + c_2 = u_1$. Solving gives

$$c_1 = \left(\frac{u_0 - u_1}{b - a}\right)ab \quad \text{and} \quad c_2 = \frac{u_1 b - u_0 a}{b - a}.$$

Thus
$$u(r) = \left(\frac{u_0 - u_1}{b - a}\right)\frac{ab}{r} + \frac{u_1 b - u_0 a}{b - a}.$$

Exercises 3.8

3. From $Dx = -y + t$ and $Dy = x - t$ we obtain $y = t - Dx$, $Dy = 1 - D^2 x$, and $(D^2 + 1)x = 1 + t$. Then

$$x = c_1 \cos t + c_2 \sin t + 1 + t$$

and

$$y = c_1 \sin t - c_2 \cos t + t - 1.$$

6. From $(D+1)x + (D-1)y = 2$ and $3x + (D+2)y = -1$ we obtain $x = -\frac{1}{3} - \frac{1}{3}(D+2)y$, $Dx = -\frac{1}{3}(D^2 + 2D)y$, and $(D^2 + 5)y = -7$. Then

$$y = c_1 \cos \sqrt{5}\, t + c_2 \sin \sqrt{5}\, t - \frac{7}{5}$$

and

$$x = \left(-\frac{2}{3}c_1 - \frac{\sqrt{5}}{3}c_2\right)\cos \sqrt{5}\, t + \left(\frac{\sqrt{5}}{3}c_1 - \frac{2}{3}c_2\right)\sin \sqrt{5}\, t + \frac{3}{5}.$$

9. From $Dx + D^2 y = e^{3t}$ and $(D+1)x + (D-1)y = 4e^{3t}$ we obtain $D(D^2 + 1)x = 34e^{3t}$ and $D(D^2 + 1)y = -8e^{3t}$. Then

$$y = c_1 + c_2 \sin t + c_3 \cos t - \frac{4}{15}e^{3t}$$

and

$$x = c_4 + c_5 \sin t + c_6 \cos t + \frac{17}{15}e^{3t}.$$

Substituting into $(D+1)x + (D-1)y = 4e^{3t}$ gives

$$(c_4 - c_1) + (c_5 - c_6 - c_3 - c_2)\sin t + (c_6 + c_5 + c_2 - c_3)\cos t = 0$$

Exercises 3.8

so that $c_4 = c_1$, $c_5 = c_3$, $c_6 = -c_2$, and

$$x = c_1 - c_2 \cos t + c_3 \sin t + \frac{17}{15}e^{3t}.$$

12. From $(2D^2-D-1)x-(2D+1)y = 1$ and $(D-1)x+Dy = -1$ we obtain $(2D+1)(D-1)(D+1)x = -1$ and $(2D+1)(D+1)y = -2$. Then

$$x = c_1 e^{-t/2} + c_2 e^{-t} + c_3 e^t + 1$$

and

$$y = c_4 e^{-t/2} + c_5 e^{-t} - 2.$$

Substituting into $(D-1)x + Dy = -1$ gives

$$\left(-\frac{3}{2}c_1 - \frac{1}{2}c_4\right)e^{-t/2} + (-2c_2 - c_5)e^{-t} = 0$$

so that $c_4 = -3c_1$, $c_5 = -2c_2$, and

$$y = -3c_1 e^{-t/2} - 2c_2 e^{-t} - 2.$$

15. Multiplying the first equation by $D+1$ and the second equation by D^2+1 and subtracting we obtain $(D^4 - D^2)x = 1$. Then

$$x = c_1 + c_2 t + c_3 e^t + c_4 e^{-t} - \frac{1}{2}t^2.$$

Multiplying the first equation by $D+1$ and subtracting we obtain $D^2(D+1)y = 1$. Then

$$y = c_5 + c_6 t + c_7 e^{-t} - \frac{1}{2}t^2.$$

Substituting into $(D-1)x + (D^2+1)y = 1$ gives

$$(-c_1 + c_2 + c_5 - 1) + (-2c_4 + 2c_7)e^{-t} + (-1 - c_2 + c_6)t = 1$$

so that $c_5 = c_1 - c_2 + 2$, $c_6 = c_2 + 1$, and $c_7 = c_4$. The solution of the system is

$$x = c_1 + c_2 t + c_3 e^t + c_4 e^{-t} - \frac{1}{2}t^2$$

$$y = (c_1 - c_2 + 2) + (c_2 + 1)t + c_4 e^{-t} - \frac{1}{2}t^2.$$

18. From $Dx + z = e^t$, $(D-1)x + Dy + Dz = 0$, and $x + 2y + Dz = e^t$ we obtain $z = -Dx + e^t$, $Dz = -D^2 x + e^t$, and the system $(-D^2 + D - 1)x + Dy = -e^t$ and $(-D^2 + 1)x + 2y = 0$. Then $y = \frac{1}{2}(D^2 - 1)x$, $Dy = \frac{1}{2}D(D^2 - 1)x$, and $(D-2)(D^2+1)x = -2e^t$ so that

$$x = c_1 e^{2t} + c_2 \cos t + c_3 \sin t + e^t,$$

$$y = \frac{3}{2}c_1 e^{2t} - c_2 \cos t - c_3 \sin t,$$

and

$$z = -2c_1e^{2t} - c_3\cos t + c_2\sin t.$$

21. From $(D+5)x+y = 0$ and $4x-(D+1)y = 0$ we obtain $y = -(D+5)x$ so that $Dy = -(D^2+5D)x$. Then $4x + (D^2+5D)x + (D+5)x = 0$ and $(D+3)^2 x = 0$. Thus

$$x = c_1 e^{-3t} + c_2 t e^{-3t}$$

and

$$y = -(2c_1 + c_2)e^{-3t} - 2c_2 t e^{-3t}.$$

Using $x(1) = 0$ and $y(1) = 1$ we obtain

$$c_1 e^{-3} + c_2 e^{-3} = 0$$

$$-(2c_1 + c_2)e^{-3} - 2c_2 e^{-3} = 1$$

or

$$c_1 + c_2 = 0$$

$$2c_1 + 3c_2 = -e^3.$$

Thus $c_1 = e^3$ and $c_2 = -e^3$. The solution of the initial value problem is

$$x = e^{-3t+3} - te^{-3t+3}$$

$$y = -e^{-3t+3} + 2te^{-3t+3}.$$

Exercises 3.9

3. The substitution $y' = u$ leads to the system

$$y' = u, \qquad u' = 4u - 4y.$$

Using formulas (5) and (6) in the text with y corresponding to x, and u corresponding to y, we obtain

Runge-Kutta method with h=0.2

m1	m2	m3	m4	k1	k2	k3	k4	t	x	y
								0.00	-2.0000	1.0000
0.2000	0.4400	0.5280	0.9072	2.4000	3.2800	3.5360	4.8064	0.20	-1.4928	4.4731

Runge-Kutta method with h=0.1

m1	m2	m3	m4	k1	k2	k3	k4	t	x	y
								0.00	-2.0000	1.0000
0.1000	0.1600	0.1710	0.2452	1.2000	1.4200	1.4520	1.7124	0.10	-1.8321	2.4427
0.2443	0.3298	0.3444	0.4487	1.7099	2.0031	2.0446	2.3900	0.20	-1.4919	4.4753

The exact value at $t = 0.2$ is $x(0.2) = -1.4918$ from Problem 1.

Exercises 3.9

6.

								Runge-Kutta method with h=0.1		
m1	m2	m3	m4	k1	k2	k3	k4	t	i1	i2
								0.00	0.0000	0.0000
10.0000	0.0000	12.5000	−20.0000	0.0000	5.0000	−5.0000	22.5000	0.10	2.5000	3.7500
8.7500	−2.5000	13.4375	−28.7500	−5.0000	4.3750	−10.6250	29.6875	0.20	2.8125	5.7813
10.1563	−4.3750	17.0703	−40.0000	−8.7500	5.0781	−16.0156	40.3516	0.30	2.0703	7.4023
13.2617	−6.3672	22.9443	−55.1758	−12.7344	6.6309	−22.5488	55.3076	0.40	0.6104	9.1919
17.9712	−8.8867	31.3507	−75.9326	−17.7734	8.9856	−31.2024	75.9821	0.50	−1.5619	11.4877

9.

								Runge-Kutta method with h=0.2		
m1	m2	m3	m4	k1	k2	k3	k4	t	x	y
								0.00	−3.0000	5.0000
−1.0000	−0.9200	−0.9080	−0.8176	−0.6000	−0.7200	−0.7120	−0.8216	0.20	−3.9123	4.2857

								Runge-Kutta method with h=0.1		
m1	m2	m3	m4	k1	k2	k3	k4	t	x	y
								0.00	−3.0000	5.0000
−0.5000	−0.4800	−0.4785	−0.4571	−0.3000	−0.3300	−0.3290	−0.3579	0.10	−3.4790	4.6707
−0.4571	−0.4342	−0.4328	−0.4086	−0.3579	−0.3858	−0.3846	−0.4112	0.20	−3.9123	4.2857

12. Solving for x' and y' we obtain the system

$$x' = \frac{1}{2}y - 3t^2 + 2t - 5$$

$$y' = -\frac{1}{2}y + 3t^2 + 2t + 5.$$

								Runge-Kutta method with h=0.2		
m1	m2	m3	m4	k1	k2	k3	k4	t	x	y
								0.00	3.0000	−1.0000
−1.1000	−1.0110	−1.0115	−0.9349	1.1000	1.0910	1.0915	1.0949	0.20	1.9867	0.0933

								Runge-Kutta method with h=0.1		
m1	m2	m3	m4	k1	k2	k3	k4	t	x	y
								0.00	3.0000	−1.0000
−0.5500	−0.5270	−0.5271	−0.5056	0.5500	0.5470	0.5471	0.5456	0.10	2.4727	−0.4527
−0.5056	−0.4857	−0.4857	−0.4673	0.5456	0.5457	0.5457	0.5473	0.20	1.9867	0.0933

Exercises 3.10

3. (a) The period corresponding to $x(0) = 1$, $x'(0) = 1$ is approximately 5.8. The second initial-value problem does not have a periodic solution.

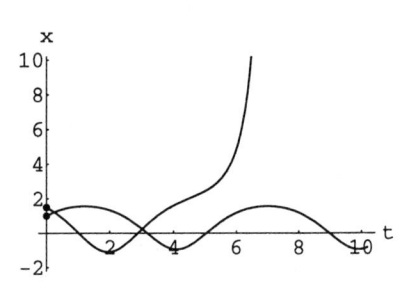

(b) The corresponding system is

$$\frac{dx}{dt} = y, \qquad \frac{dy}{dt} = x^2 - 2x.$$

Critical points are $(0,0)$ and $(2,0)$.

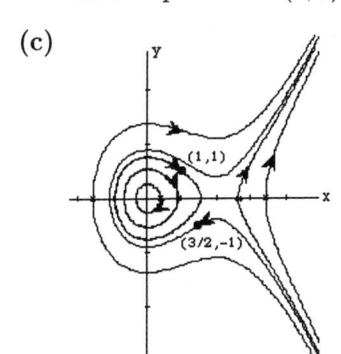

(c)

$(0,0)$ is a stable center; $(2,0)$ is an unstable saddle point.

6. (a) From the graph we see that $a \leq x \leq 1$, where $a \approx -0.73$.

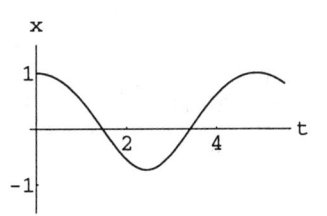

(b) From the system

$$\frac{dx}{dt} = y, \qquad \frac{dy}{dt} = -2x + x^2$$

we obtain

$$\frac{dy}{dx} = \frac{-2x + x^2}{y}.$$

Exercises 3.10

Using separation of variables we obtain $y^2 = -2x^2 + \frac{2}{3}x^3 + \frac{4}{3}$. At the extreme displacement the velocity of the mass is 0; that is, $y = 0$. Solving $2x^3 - 6x^2 + 4 = 0$ we obtain $x = 1$ and $x = 1 \pm \sqrt{3}$. Since we want x to be negative, we take $x = 1 - \sqrt{3} \approx -0.73$.

9. (a) This is a damped hard spring, so all solutions should be oscillatory with $x \to 0$ as $t \to \infty$.

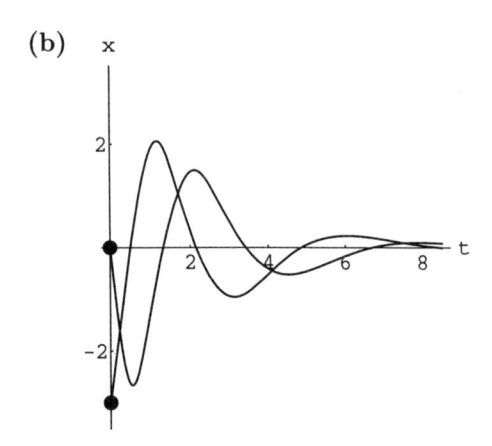

   (b)

   (c) The only critical point is $(0,0)$.

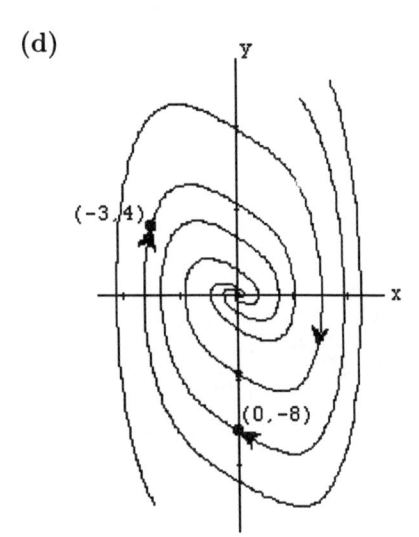

   (d)

   $(0,0)$ is an asymptotically stable spiral point.

12. (a)

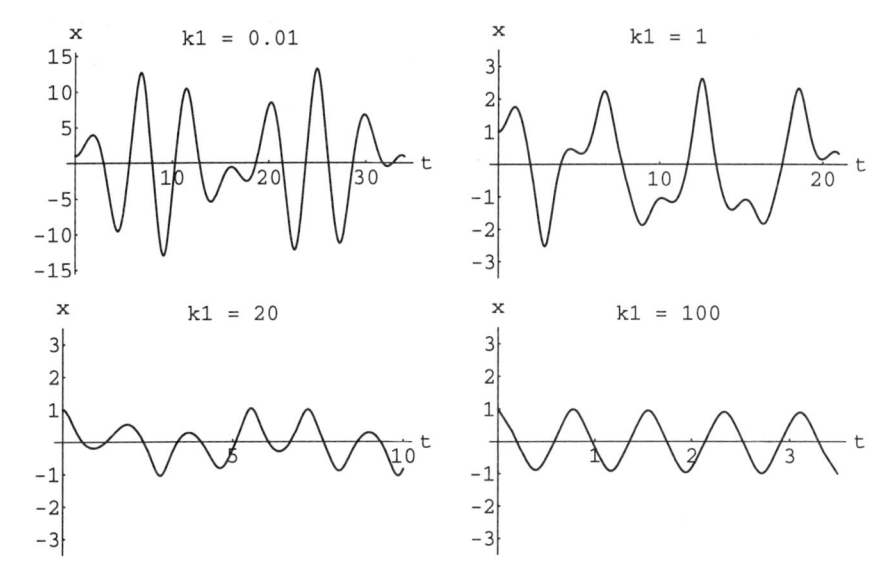

[Plots for k1 = 0.01, k1 = 1, k1 = 20, k1 = 100]

When k_1 is very small the effect of the nonlinearity is greatly diminished, and the system is close to pure resonance.

(b)

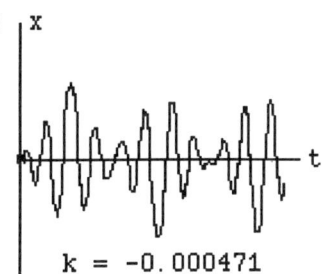

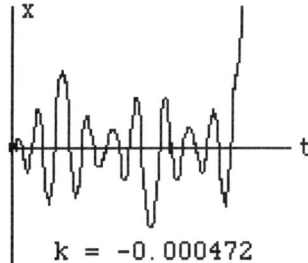

[Plots for k = -0.000471 and k = -0.000472]

The system appears to be oscillatory for $-0.000471 \le k_1 < 0$ and nonoscillatory for $k_1 \le -0.000472$.

(c)

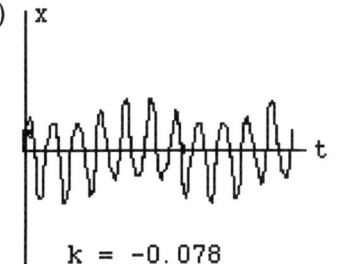

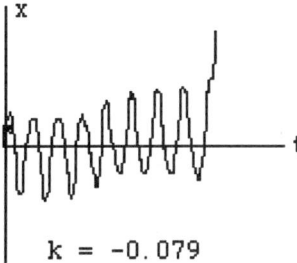

[Plots for k = -0.078 and k = -0.079]

The system appears to be oscillatory for $-0.077 \le k_1 < 0$ and nonoscillatory for $k_1 \le 0.078$.

Exercises 3.10

15. Let $u = y'$ so that $y'' = u\dfrac{du}{dy}$. The equation becomes $yu\dfrac{du}{dy} = u^2$. Separating variables we obtain

$$\dfrac{du}{u} = \dfrac{dy}{y} \implies \ln|u| = \ln|y| + c_1 \implies u = c_2 y$$

$$\implies \dfrac{dy}{dt} = c_2 y \implies \dfrac{dy}{y} = c_2\,dt \implies \ln|y| = c_2 t + c_3$$

$$\implies y = c_4 e^{c_2 t}.$$

18. Let $u = \dfrac{dx}{dt}$ so that $\dfrac{d^2x}{dt^2} = u\dfrac{du}{dx}$. The equation becomes $u\dfrac{du}{dx} = \dfrac{-k^2}{x^2}$. Separating variables we obtain

$$u\,du = -\dfrac{k^2}{x^2}\,dx \implies \dfrac{1}{2}u^2 = \dfrac{k^2}{x} + c \implies \dfrac{1}{2}v^2 = \dfrac{k^2}{x} + c.$$

When $t = 0$, $x = x_0$ and $v = 0$ so $0 = \dfrac{k^2}{x_0} + c$ and $c = -\dfrac{k^2}{x_0}$. Then

$$\dfrac{1}{2}v^2 = k^2\left(\dfrac{1}{x} - \dfrac{1}{x_0}\right) \quad \text{and} \quad \dfrac{dx}{dt} = -k\sqrt{2}\sqrt{\dfrac{x_0 - x}{xx_0}}.$$

Separating variables we have

$$-\sqrt{\dfrac{xx_0}{x_0 - x}}\,dx = k\sqrt{2}\,dt \implies t = -\dfrac{1}{k}\sqrt{\dfrac{x_0}{2}}\int\sqrt{\dfrac{x}{x_0 - x}}\,dx.$$

Using *Mathematica* to integrate we obtain

$$t = -\dfrac{1}{k}\sqrt{\dfrac{x_0}{2}}\left[-\sqrt{x(x_0 - x)} - \dfrac{x_0}{2}\tan^{-1}\dfrac{(x_0 - 2x)}{2x}\sqrt{\dfrac{x}{x_0 - x}}\right]$$

$$= \dfrac{1}{k}\sqrt{\dfrac{x_0}{2}}\left[\sqrt{x(x_0 - x)} + \dfrac{x_0}{2}\tan^{-1}\dfrac{x_0 - 2x}{2\sqrt{x(x_0 - x)}}\right].$$

4 Systems of First-Order Equations

Exercises 4.1

3. Let $\mathbf{X} = \begin{pmatrix} x \\ y \\ z \end{pmatrix}$. Then

$$\mathbf{X}' = \begin{pmatrix} -3 & 4 & -9 \\ 6 & -1 & 0 \\ 10 & 4 & 3 \end{pmatrix} \mathbf{X}.$$

6. Let $\mathbf{X} = \begin{pmatrix} x \\ y \end{pmatrix}$. Then

$$\mathbf{X}' = \begin{pmatrix} -3 & 4 \\ 5 & 9 \end{pmatrix} \mathbf{X} + \begin{pmatrix} e^{-t} \sin 2t \\ 4e^{-t} \cos 2t \end{pmatrix}.$$

9. $\dfrac{dx}{dt} = x - y + 2z + e^{-t} - 3t; \quad \dfrac{dy}{dt} = 3x - 4y + z + 2e^{-t} + t; \quad \dfrac{dz}{dt} = -2x + 5y + 6z + 2e^{-t} - t$

12. Since

$$\mathbf{X}' = \begin{pmatrix} 5 \cos t - 5 \sin t \\ 2 \cos t - 4 \sin t \end{pmatrix} e^t \quad \text{and} \quad \begin{pmatrix} -2 & 5 \\ -2 & 4 \end{pmatrix} \mathbf{X} = \begin{pmatrix} 5 \cos t - 5 \sin t \\ 2 \cos t - 4 \sin t \end{pmatrix} e^t$$

we see that

$$\mathbf{X}' = \begin{pmatrix} -2 & 5 \\ -2 & 4 \end{pmatrix} \mathbf{X}.$$

15. Since

$$\mathbf{X}' = \begin{pmatrix} 0 \\ 0 \\ 0 \end{pmatrix} \quad \text{and} \quad \begin{pmatrix} 1 & 2 & 1 \\ 6 & -1 & 0 \\ -1 & -2 & -1 \end{pmatrix} \mathbf{X} = \begin{pmatrix} 0 \\ 0 \\ 0 \end{pmatrix}$$

we see that

$$\mathbf{X}' = \begin{pmatrix} 1 & 2 & 1 \\ 6 & -1 & 0 \\ -1 & -2 & -1 \end{pmatrix} \mathbf{X}.$$

18. Yes, since $W(\mathbf{X}_1, \mathbf{X}_2) = 8e^{2t} \neq 0$ and $\mathbf{X}_1$ and $\mathbf{X}_2$ are linearly independent on $-\infty < t < \infty$.

21. Since

$$\mathbf{X}'_p = \begin{pmatrix} 2 \\ -1 \end{pmatrix} \quad \text{and} \quad \begin{pmatrix} 1 & 4 \\ 3 & 2 \end{pmatrix} \mathbf{X}_p + \begin{pmatrix} 2 \\ -4 \end{pmatrix} t + \begin{pmatrix} -7 \\ -18 \end{pmatrix} = \begin{pmatrix} 2 \\ -1 \end{pmatrix}$$

we see that

$$\mathbf{X}'_p = \begin{pmatrix} 1 & 4 \\ 3 & 2 \end{pmatrix} \mathbf{X}_p + \begin{pmatrix} 2 \\ -4 \end{pmatrix} t + \begin{pmatrix} -7 \\ -18 \end{pmatrix}.$$

Exercises 4.1

24. Since

$$\mathbf{X}'_p = \begin{pmatrix} 3\cos 3t \\ 0 \\ -3\sin 3t \end{pmatrix} \text{ and } \begin{pmatrix} 1 & 2 & 3 \\ -4 & 2 & 0 \\ -6 & 1 & 0 \end{pmatrix} \mathbf{X}_p + \begin{pmatrix} -1 \\ 4 \\ 3 \end{pmatrix} \sin 3t = \begin{pmatrix} 3\cos 3t \\ 0 \\ -3\sin 3t \end{pmatrix}$$

we see that

$$\mathbf{X}'_p = \begin{pmatrix} 1 & 2 & 3 \\ -4 & 2 & 0 \\ -6 & 1 & 0 \end{pmatrix} \mathbf{X}_p + \begin{pmatrix} -1 \\ 4 \\ 3 \end{pmatrix} \sin 3t.$$

Exercises 4.2

3. The system is

$$\mathbf{X}' = \begin{pmatrix} -4 & 2 \\ -5/2 & 2 \end{pmatrix} \mathbf{X}$$

and $\det(\mathbf{A} - \lambda \mathbf{I}) = (\lambda - 1)(\lambda + 3) = 0$. For $\lambda_1 = 1$ we obtain

$$\begin{pmatrix} -5 & 2 & | & 0 \\ -5/2 & 1 & | & 0 \end{pmatrix} \implies \begin{pmatrix} -5 & 2 & | & 0 \\ 0 & 0 & | & 0 \end{pmatrix} \text{ so that } \mathbf{K}_1 = \begin{pmatrix} 2 \\ 5 \end{pmatrix}.$$

For $\lambda_2 = -3$ we obtain

$$\begin{pmatrix} -1 & 2 & | & 0 \\ -5/2 & 5 & | & 0 \end{pmatrix} \implies \begin{pmatrix} -1 & 2 & | & 0 \\ 0 & 0 & | & 0 \end{pmatrix} \text{ so that } \mathbf{K}_2 = \begin{pmatrix} 2 \\ 1 \end{pmatrix}.$$

Then

$$\mathbf{X} = c_1 \begin{pmatrix} 2 \\ 5 \end{pmatrix} e^t + c_2 \begin{pmatrix} 2 \\ 1 \end{pmatrix} e^{-3t}.$$

6. The system is

$$\mathbf{X}' = \begin{pmatrix} -6 & 2 \\ -3 & 1 \end{pmatrix} \mathbf{X}$$

and $\det(\mathbf{A} - \lambda \mathbf{I}) = \lambda(\lambda + 5) = 0$. For $\lambda_1 = 0$ we obtain

$$\begin{pmatrix} -6 & 2 & | & 0 \\ -3 & 1 & | & 0 \end{pmatrix} \implies \begin{pmatrix} 1 & -1/3 & | & 0 \\ 0 & 0 & | & 0 \end{pmatrix} \text{ so that } \mathbf{K}_1 = \begin{pmatrix} 1 \\ 3 \end{pmatrix}.$$

For $\lambda_2 = -5$ we obtain

$$\begin{pmatrix} -1 & 2 & | & 0 \\ -3 & 6 & | & 0 \end{pmatrix} \implies \begin{pmatrix} 1 & -2 & | & 0 \\ 0 & 0 & | & 0 \end{pmatrix} \text{ so that } \mathbf{K}_2 = \begin{pmatrix} 2 \\ 1 \end{pmatrix}.$$

Then

$$\mathbf{X} = c_1 \begin{pmatrix} 1 \\ 3 \end{pmatrix} + c_2 \begin{pmatrix} 2 \\ 1 \end{pmatrix} e^{-5t}.$$

Exercises 4.2

9. We have $\det(\mathbf{A} - \lambda\mathbf{I}) = -(\lambda+1)(\lambda-3)(\lambda+2) = 0$. For $\lambda_1 = -1$, $\lambda_2 = 3$, and $\lambda_3 = -2$ we obtain

$$\mathbf{K}_1 = \begin{pmatrix} -1 \\ 0 \\ 1 \end{pmatrix}, \quad \mathbf{K}_2 = \begin{pmatrix} 1 \\ 4 \\ 3 \end{pmatrix}, \quad \text{and} \quad \mathbf{K}_3 = \begin{pmatrix} 1 \\ -1 \\ 3 \end{pmatrix},$$

so that

$$\mathbf{X} = c_1 \begin{pmatrix} -1 \\ 0 \\ 1 \end{pmatrix} e^{-t} + c_2 \begin{pmatrix} 1 \\ 4 \\ 3 \end{pmatrix} e^{3t} + c_3 \begin{pmatrix} 1 \\ -1 \\ 3 \end{pmatrix} e^{-2t}.$$

12. We have $\det(\mathbf{A} - \lambda\mathbf{I}) = (\lambda-3)(\lambda+5)(6-\lambda) = 0$. For $\lambda_1 = 3$, $\lambda_2 = -5$, and $\lambda_3 = 6$ we obtain

$$\mathbf{K}_1 = \begin{pmatrix} 1 \\ 1 \\ 0 \end{pmatrix}, \quad \mathbf{K}_2 = \begin{pmatrix} 1 \\ -1 \\ 0 \end{pmatrix}, \quad \text{and} \quad \mathbf{K}_3 = \begin{pmatrix} 2 \\ -2 \\ 11 \end{pmatrix},$$

so that

$$\mathbf{X} = c_1 \begin{pmatrix} 1 \\ 1 \\ 0 \end{pmatrix} e^{3t} + c_2 \begin{pmatrix} 1 \\ -1 \\ 0 \end{pmatrix} e^{-5t} + c_3 \begin{pmatrix} 2 \\ -2 \\ 11 \end{pmatrix} e^{6t}.$$

15. $\mathbf{X} = c_1 \begin{pmatrix} 0.382175 \\ 0.851161 \\ 0.359815 \end{pmatrix} e^{8.58979t} + c_2 \begin{pmatrix} 0.405188 \\ -0.676043 \\ 0.615458 \end{pmatrix} e^{2.25684t} + c_3 \begin{pmatrix} -0.923562 \\ -0.132174 \\ 0.35995 \end{pmatrix} e^{-0.0466321t}.$

18. We have $\det(\mathbf{A} - \lambda\mathbf{I}) = (\lambda+1)^2 = 0$. For $\lambda_1 = -1$ we obtain

$$\mathbf{K} = \begin{pmatrix} 1 \\ 1 \end{pmatrix}.$$

A solution of $(\mathbf{A} - \lambda_1\mathbf{I})\mathbf{P} = \mathbf{K}$ is

$$\mathbf{P} = \begin{pmatrix} 0 \\ 1/5 \end{pmatrix}$$

so that

$$\mathbf{X} = c_1 \begin{pmatrix} 1 \\ 1 \end{pmatrix} e^{-t} + c_2 \left[\begin{pmatrix} 1 \\ 1 \end{pmatrix} te^{-t} + \begin{pmatrix} 0 \\ 1/5 \end{pmatrix} e^{-t} \right].$$

21. We have $\det(\mathbf{A} - \lambda\mathbf{I}) = (1-\lambda)(\lambda-2)^2 = 0$. For $\lambda_1 = 1$ we obtain

$$\mathbf{K}_1 = \begin{pmatrix} 1 \\ 1 \\ 1 \end{pmatrix}.$$

Exercises 4.2

For $\lambda_2 = 2$ we obtain

$$K_2 = \begin{pmatrix} 1 \\ 0 \\ 1 \end{pmatrix} \quad \text{and} \quad K_3 = \begin{pmatrix} 1 \\ 1 \\ 0 \end{pmatrix}.$$

Then

$$X = c_1 \begin{pmatrix} 1 \\ 1 \\ 1 \end{pmatrix} e^t + c_2 \begin{pmatrix} 1 \\ 0 \\ 1 \end{pmatrix} e^{2t} + c_3 \begin{pmatrix} 1 \\ 1 \\ 0 \end{pmatrix} e^{2t}.$$

24. We have $\det(A - \lambda I) = (1 - \lambda)(\lambda - 2)^2 = 0$. For $\lambda_1 = 1$ we obtain

$$K_1 = \begin{pmatrix} 1 \\ 0 \\ 0 \end{pmatrix}.$$

For $\lambda_2 = 2$ we obtain

$$K = \begin{pmatrix} 0 \\ -1 \\ 1 \end{pmatrix}.$$

A solution of $(A - \lambda_2 I)P = K$ is

$$P = \begin{pmatrix} 0 \\ -1 \\ 0 \end{pmatrix}$$

so that

$$X = c_1 \begin{pmatrix} 1 \\ 0 \\ 0 \end{pmatrix} e^t + c_2 \begin{pmatrix} 0 \\ -1 \\ 1 \end{pmatrix} e^{2t} + c_3 \left[\begin{pmatrix} 0 \\ -1 \\ 1 \end{pmatrix} te^{2t} + \begin{pmatrix} 0 \\ -1 \\ 0 \end{pmatrix} e^{2t} \right].$$

27. We have $\det(A - \lambda I) = (\lambda - 4)^2 = 0$. For $\lambda_1 = 4$ we obtain

$$K = \begin{pmatrix} 2 \\ 1 \end{pmatrix}.$$

A solution of $(A - \lambda_1 I)P = K$ is

$$P = \begin{pmatrix} 1 \\ 1 \end{pmatrix}$$

so that

$$X = c_1 \begin{pmatrix} 2 \\ 1 \end{pmatrix} e^{4t} + c_2 \left[\begin{pmatrix} 2 \\ 1 \end{pmatrix} te^{4t} + \begin{pmatrix} 1 \\ 1 \end{pmatrix} e^{4t} \right].$$

If

$$X(0) = \begin{pmatrix} -1 \\ 6 \end{pmatrix}$$

Exercises 4.2

then $c_1 = -7$ and $c_2 = 13$.

30. The system of differential equations is

$$x_1' = 2x_1 + x_2$$
$$x_2' = 2x_2$$
$$x_3' = 2x_3$$
$$x_4' = 2x_4 + x_5$$
$$x_5' = 2x_5.$$

We see immediately that $x_2 = c_2 e^{2t}$, $x_3 = c_3 e^{2t}$, and $x_5 = c_5 e^{2t}$. Then

$$x_1' = 2x_1 + c_2 e^{2t} \implies x_1 = c_2 t e^{2t} + c_1 e^{2t}$$

and

$$x_4' = 2x_4 + c_5 e^{2t} \implies x_4 = c_5 t e^{2t} + c_4 e^{2t}.$$

The general solution of the system is

$$\mathbf{X} = \begin{pmatrix} c_2 t e^{2t} + c_1 e^{2t} \\ c_2 e^{2t} \\ c_3 e^{2t} \\ c_5 t e^{2t} + c_4 e^{2t} \\ c_5 e^{2t} \end{pmatrix} = c_1 \begin{pmatrix} 1 \\ 0 \\ 0 \\ 0 \\ 0 \end{pmatrix} e^{2t} + c_2 \left[\begin{pmatrix} 1 \\ 0 \\ 0 \\ 0 \\ 0 \end{pmatrix} t e^{2t} + \begin{pmatrix} 0 \\ 1 \\ 0 \\ 0 \\ 0 \end{pmatrix} e^{2t} \right]$$

$$+ c_3 \begin{pmatrix} 0 \\ 0 \\ 1 \\ 0 \\ 0 \end{pmatrix} e^{2t} + c_4 \begin{pmatrix} 0 \\ 0 \\ 0 \\ 1 \\ 0 \end{pmatrix} e^{2t} + c_5 \left[\begin{pmatrix} 0 \\ 0 \\ 0 \\ 1 \\ 0 \end{pmatrix} t e^{2t} + \begin{pmatrix} 0 \\ 0 \\ 0 \\ 0 \\ 1 \end{pmatrix} e^{2t} \right]$$

$$= c_1 \mathbf{K}_1 e^{2t} + c_2 \left[\mathbf{K}_1 t e^{2t} + \begin{pmatrix} 0 \\ 1 \\ 0 \\ 0 \\ 0 \end{pmatrix} e^{2t} \right] + c_3 \mathbf{K}_2 e^{2t} + c_4 \mathbf{K}_3 e^{2t} + c_5 \left[\mathbf{K}_3 t e^{2t} + \begin{pmatrix} 0 \\ 0 \\ 0 \\ 0 \\ 1 \end{pmatrix} e^{2t} \right].$$

Exercises 4.2

In Problems 33 and 36 the form of the answer will vary according to the choice of eigenvector. For example, in Problem 33, if $\mathbf{K}_1$ is chosen to be $\begin{pmatrix} 1 \\ -1+i \end{pmatrix}$ the solution has the form

$$\mathbf{X} = c_1 \begin{pmatrix} \cos t \\ -\cos t - \sin t \end{pmatrix} e^{4t} + c_2 \begin{pmatrix} \sin t \\ \cos t - \sin t \end{pmatrix} e^{4t}.$$

33. We have $\det(\mathbf{A} - \lambda\mathbf{I}) = \lambda^2 - 8\lambda + 17 = 0$. For $\lambda_1 = 4 + i$ we obtain

$$\mathbf{K}_1 = \begin{pmatrix} -1-i \\ 2 \end{pmatrix}$$

so that $\quad \mathbf{X}_1 = \begin{pmatrix} -1-i \\ 2 \end{pmatrix} e^{(4+i)t} = \begin{pmatrix} \sin t - \cos t \\ 2\cos t \end{pmatrix} e^{4t} + i \begin{pmatrix} -\sin t - \cos t \\ 2\sin t \end{pmatrix} e^{4t}.$

Then $\quad \mathbf{X} = c_1 \begin{pmatrix} \sin t - \cos t \\ 2\cos t \end{pmatrix} e^{4t} + c_2 \begin{pmatrix} -\sin t - \cos t \\ 2\sin t \end{pmatrix} e^{4t}.$

36. We have $\det(\mathbf{A} - \lambda\mathbf{I}) = \lambda^2 + 2\lambda + 5 = 0$. For $\lambda_1 = -1 + 2i$ we obtain

$$\mathbf{K}_1 = \begin{pmatrix} 2+2i \\ 1 \end{pmatrix}$$

so that $\quad \mathbf{X}_1 = \begin{pmatrix} 2+2i \\ 1 \end{pmatrix} e^{(-1+2i)t}$

$= (2\cos 2t - 2\sin 2t \cos 2t) e^{-t} + i \begin{pmatrix} 2\cos 2t + 2\sin 2t \\ \sin 2t \end{pmatrix} e^{-t}.$

Then $\quad \mathbf{X} = c_1 \begin{pmatrix} 2\cos 2t - 2\sin 2t \\ \cos 2t \end{pmatrix} e^{-t} + c_2 \begin{pmatrix} 2\cos 2t + 2\sin 2t \\ \sin 2t \end{pmatrix} e^{-t}.$

39. We have $\det(\mathbf{A} - \lambda\mathbf{I}) = (1 - \lambda)(\lambda^2 - 2\lambda + 2) = 0$. For $\lambda_1 = 1$ we obtain

$$\mathbf{K}_1 = \begin{pmatrix} 0 \\ 2 \\ 1 \end{pmatrix}.$$

For $\lambda_2 = 1 + i$ we obtain

$$\mathbf{K}_2 = \begin{pmatrix} 1 \\ i \\ i \end{pmatrix}$$

46

so that
$$\mathbf{X}_2 = \begin{pmatrix} 1 \\ i \\ i \end{pmatrix} e^{(1+i)t} = \begin{pmatrix} \cos t \\ -\sin t \\ -\sin t \end{pmatrix} e^t + i \begin{pmatrix} \sin t \\ \cos t \\ \cos t \end{pmatrix} e^t.$$

Then
$$\mathbf{X} = c_1 \begin{pmatrix} 0 \\ 2 \\ 1 \end{pmatrix} e^t + c_2 \begin{pmatrix} \cos t \\ -\sin t \\ -\sin t \end{pmatrix} e^t + c_3 \begin{pmatrix} \sin t \\ \cos t \\ \cos t \end{pmatrix} e^t.$$

42. We have $\det(\mathbf{A} - \lambda \mathbf{I}) = -(\lambda+2)(\lambda^2+4) = 0$. For $\lambda_1 = -2$ we obtain
$$\mathbf{K}_1 = \begin{pmatrix} 0 \\ -1 \\ 1 \end{pmatrix}.$$

For $\lambda_2 = 2i$ we obtain
$$\mathbf{K}_2 = \begin{pmatrix} -2 - 2i \\ 1 \\ 1 \end{pmatrix}$$

so that
$$\mathbf{X}_2 = \begin{pmatrix} -2 - 2i \\ 1 \\ 1 \end{pmatrix} e^{2it} = \begin{pmatrix} -2\cos 2t + 2\sin 2t \\ \cos 2t \\ \cos 2t \end{pmatrix} + i \begin{pmatrix} -2\cos 2t - 2\sin 2t \\ \sin 2t \\ \sin 2t \end{pmatrix}.$$

Then
$$\mathbf{X} = c_1 \begin{pmatrix} 0 \\ -1 \\ 1 \end{pmatrix} e^{-2t} + c_2 \begin{pmatrix} -2\cos 2t + 2\sin 2t \\ \cos 2t \\ \cos 2t \end{pmatrix} + c_3 \begin{pmatrix} -2\cos 2t - 2\sin 2t \\ \sin 2t \\ \sin 2t \end{pmatrix}.$$

Exercises 4.3

3. From
$$\mathbf{X}' = \begin{pmatrix} 3 & -5 \\ 3/4 & -1 \end{pmatrix} \mathbf{X} + \begin{pmatrix} 1 \\ -1 \end{pmatrix} e^{t/2}$$

we obtain
$$\mathbf{X}_c = c_1 \begin{pmatrix} 10 \\ 3 \end{pmatrix} e^{3t/2} + c_2 \begin{pmatrix} 2 \\ 1 \end{pmatrix} e^{t/2}.$$

Then
$$\mathbf{\Phi} = \begin{pmatrix} 10e^{3t/2} & 2e^{t/2} \\ 3e^{3t/2} & e^{t/2} \end{pmatrix} \quad \text{and} \quad \mathbf{\Phi}^{-1} = \begin{pmatrix} \tfrac{1}{4}e^{-3t/2} & -\tfrac{1}{2}e^{-3t/2} \\ -\tfrac{3}{4}e^{-t/2} & \tfrac{5}{2}e^{-t/2} \end{pmatrix}$$

so that
$$\mathbf{U} = \int \mathbf{\Phi}^{-1} \mathbf{F}\, dt = \int \begin{pmatrix} \tfrac{3}{4}e^{-t} \\ -\tfrac{13}{4} \end{pmatrix} dt = \begin{pmatrix} -\tfrac{3}{4}e^{-t} \\ -\tfrac{13}{4}t \end{pmatrix}$$

Exercises 4.3

and
$$\mathbf{X}_p = \mathbf{\Phi U} = \begin{pmatrix} -13/2 \\ -13/4 \end{pmatrix} te^{t/2} + \begin{pmatrix} -15/2 \\ -9/4 \end{pmatrix} e^{t/2}.$$

6. From
$$\mathbf{X}' = \begin{pmatrix} 0 & 2 \\ -1 & 3 \end{pmatrix} \mathbf{X} + \begin{pmatrix} 2 \\ e^{-3t} \end{pmatrix}$$

we obtain
$$\mathbf{X}_c = c_1 \begin{pmatrix} 2 \\ 1 \end{pmatrix} e^t + c_2 \begin{pmatrix} 1 \\ 1 \end{pmatrix} e^{2t}.$$

Then
$$\mathbf{\Phi} = \begin{pmatrix} 2e^t & e^{2t} \\ e^t & e^{2t} \end{pmatrix} \quad \text{and} \quad \mathbf{\Phi}^{-1} = \begin{pmatrix} e^{-t} & -e^{-t} \\ -e^{-2t} & 2e^{-2t} \end{pmatrix}$$

so that
$$\mathbf{U} = \int \mathbf{\Phi}^{-1} \mathbf{F}\, dt = \int \begin{pmatrix} 2e^{-t} - e^{-4t} \\ -2e^{-2t} + 2e^{-5t} \end{pmatrix} dt = \begin{pmatrix} -2e^{-t} + \frac{1}{4}e^{-4t} \\ e^{-2t} - \frac{2}{5}e^{-5t} \end{pmatrix}$$

and
$$\mathbf{X}_p = \mathbf{\Phi U} = \begin{pmatrix} \frac{1}{10}e^{-3t} - 3 \\ -\frac{3}{20}e^{-3t} - 1 \end{pmatrix}.$$

9. From
$$\mathbf{X}' = \begin{pmatrix} 3 & 2 \\ -2 & -1 \end{pmatrix} \mathbf{X} + \begin{pmatrix} 2 \\ 1 \end{pmatrix} e^{-t}$$

we obtain
$$\mathbf{X}_c = c_1 \begin{pmatrix} 1 \\ -1 \end{pmatrix} e^t + c_2 \left[\begin{pmatrix} 1 \\ -1 \end{pmatrix} te^t + \begin{pmatrix} 0 \\ 1/2 \end{pmatrix} e^t \right].$$

Then
$$\mathbf{\Phi} = \begin{pmatrix} e^t & te^t \\ -e^t & \frac{1}{2}e^t - te^t \end{pmatrix} \quad \text{and} \quad \mathbf{\Phi}^{-1} = \begin{pmatrix} e^{-t} - 2te^{-t} & -2te^{-t} \\ 2e^{-t} & 2e^{-t} \end{pmatrix}$$

so that
$$\mathbf{U} = \int \mathbf{\Phi}^{-1} \mathbf{F}\, dt = \int \begin{pmatrix} 2e^{-2t} - 6te^{-2t} \\ 6e^{-2t} \end{pmatrix} dt = \begin{pmatrix} \frac{1}{2}e^{-2t} + 3te^{-2t} \\ -3e^{-2t} \end{pmatrix}$$

and
$$\mathbf{X}_p = \mathbf{\Phi U} = \begin{pmatrix} 1/2 \\ -2 \end{pmatrix} e^{-t}.$$

12. From
$$\mathbf{X}' = \begin{pmatrix} 1 & -1 \\ 1 & 1 \end{pmatrix} \mathbf{X} + \begin{pmatrix} 3 \\ 3 \end{pmatrix} e^t$$

we obtain
$$\mathbf{X}_c = c_1 \begin{pmatrix} -\sin t \\ \cos t \end{pmatrix} e^t + c_2 \begin{pmatrix} \cos t \\ \sin t \end{pmatrix} e^t.$$

Then
$$\mathbf{\Phi} = \begin{pmatrix} -\sin t & \cos t \\ \cos t & \sin t \end{pmatrix} e^t \quad \text{and} \quad \mathbf{\Phi}^{-1} = \begin{pmatrix} -\sin t & \cos t \\ \cos t & \sin t \end{pmatrix} e^{-t}$$

48

Exercises 4.3

so that
$$U = \int \Phi^{-1}F\, dt = \int \begin{pmatrix} -3\sin t + 3\cos t \\ 3\cos t + 3\sin t \end{pmatrix} dt = \begin{pmatrix} 3\cos t + 3\sin t \\ 3\sin t - 3\cos t \end{pmatrix}$$

and
$$X_p = \Phi U = \begin{pmatrix} -3 \\ 3 \end{pmatrix} e^t.$$

15. From
$$X' = \begin{pmatrix} 0 & 1 \\ -1 & 0 \end{pmatrix} X + \begin{pmatrix} 0 \\ \sec t \tan t \end{pmatrix}$$

we obtain
$$X_c = c_1 \begin{pmatrix} \cos t \\ -\sin t \end{pmatrix} + c_2 \begin{pmatrix} \sin t \\ \cos t \end{pmatrix}.$$

Then
$$\Phi = \begin{pmatrix} \cos t & \sin t \\ -\sin t & \cos t \end{pmatrix} t \quad \text{and} \quad \Phi^{-1} = \begin{pmatrix} \cos t & -\sin t \\ \sin t & \cos t \end{pmatrix}$$

so that
$$U = \int \Phi^{-1}F\, dt = \int \begin{pmatrix} -\tan^2 t \\ \tan t \end{pmatrix} dt = \begin{pmatrix} t - \tan t \\ \ln|\sec t| \end{pmatrix}$$

and
$$X_p = \Phi U = \begin{pmatrix} \cos t \\ -\sin t \end{pmatrix} t + \begin{pmatrix} -\sin t \\ \sin t \tan t \end{pmatrix} + \begin{pmatrix} \sin t \\ \cos t \end{pmatrix} \ln|\sec t|.$$

18. From
$$X' = \begin{pmatrix} 1 & -2 \\ 1 & -1 \end{pmatrix} X + \begin{pmatrix} \tan t \\ 1 \end{pmatrix}$$

we obtain
$$X_c = c_1 \begin{pmatrix} \cos t - \sin t \\ \cos t \end{pmatrix} + c_2 \begin{pmatrix} \cos t + \sin t \\ \sin t \end{pmatrix}.$$

Then
$$\Phi = \begin{pmatrix} \cos t - \sin t & \cos t + \sin t \\ \cos t & \sin t \end{pmatrix} \quad \text{and} \quad \Phi^{-1} = \begin{pmatrix} -\sin t & \cos t + \sin t \\ \cos t & \sin t - \cos t \end{pmatrix}$$

so that
$$U = \int \Phi^{-1}F\, dt = \int \begin{pmatrix} 2\cos t + \sin t - \sec t \\ 2\sin t - \cos t \end{pmatrix} dt = \begin{pmatrix} 2\sin t - \cos t - \ln|\sec t + \tan t| \\ -2\cos t - \sin t \end{pmatrix}$$

and
$$X_p = \Phi U = \begin{pmatrix} 3\sin t \cos t - \cos^2 t - 2\sin^2 t + (\sin t - \cos t)\ln|\sec t + \tan t| \\ \sin^2 t - \cos^2 t - \cos t(\ln|\sec t + \tan t|) \end{pmatrix}.$$

21. From
$$X' = \begin{pmatrix} 3 & -1 \\ -1 & 3 \end{pmatrix} X + \begin{pmatrix} 4e^{2t} \\ 4e^{4t} \end{pmatrix}$$

we obtain
$$\Phi = \begin{pmatrix} -e^{4t} & e^{2t} \\ e^{4t} & e^{2t} \end{pmatrix}, \quad \Phi^{-1} = \begin{pmatrix} -\tfrac{1}{2}e^{-4t} & \tfrac{1}{2}e^{-4t} \\ \tfrac{1}{2}e^{-2t} & \tfrac{1}{2}e^{-2t} \end{pmatrix},$$

49

Exercises 4.3

and
$$X = \Phi\Phi^{-1}(0)X(0) + \Phi \int_0^t \Phi^{-1}F\, ds = \Phi \cdot \begin{pmatrix} 0 \\ 1 \end{pmatrix} + \Phi \cdot \begin{pmatrix} e^{-2t} + 2t - 1 \\ e^{2t} + 2t - 1 \end{pmatrix}$$

$$= \begin{pmatrix} 2 \\ 2 \end{pmatrix} te^{2t} + \begin{pmatrix} -1 \\ 1 \end{pmatrix} e^{2t} + \begin{pmatrix} -2 \\ 2 \end{pmatrix} te^{4t} + \begin{pmatrix} 2 \\ 0 \end{pmatrix} e^{4t}.$$

Exercises 4.4

3. The system is
$$X' = \begin{pmatrix} -\frac{2}{25} & \frac{1}{50} \\ \frac{2}{25} & -\frac{2}{25} \end{pmatrix} X, \qquad X(0) = \begin{pmatrix} 25 \\ 0 \end{pmatrix},$$

and $\det(A - \lambda I) = \frac{1}{625}(25\lambda+3)(25\lambda+1) = 0$. The eigenvalues are $-\frac{3}{25}$ and $-\frac{1}{25}$ with corresponding eigenvectors

$$\begin{pmatrix} -\frac{1}{2} \\ 1 \end{pmatrix} \quad \text{and} \quad \begin{pmatrix} \frac{1}{2} \\ 1 \end{pmatrix}.$$

The general solution of the system is
$$X(t) = c_1 \begin{pmatrix} -\frac{1}{2} \\ 1 \end{pmatrix} e^{-3t/25} + c_2 \begin{pmatrix} \frac{1}{2} \\ 1 \end{pmatrix} e^{-t/25}.$$

The initial condition implies
$$c_1 \begin{pmatrix} -\frac{1}{2} \\ 1 \end{pmatrix} + c_2 \begin{pmatrix} \frac{1}{2} \\ 1 \end{pmatrix} = \begin{pmatrix} 25 \\ 0 \end{pmatrix}$$

from which we find $c_1 = -25$ and $c_2 = 25$. The solution of the initial-value problem is
$$X(t) = \begin{pmatrix} \frac{25}{2} \\ -25 \end{pmatrix} e^{-3t/25} + \begin{pmatrix} \frac{25}{2} \\ 25 \end{pmatrix} e^{-t/25}$$

or
$$x_1(t) = \frac{25}{2} e^{-3t/25} + \frac{25}{2} e^{-t/25}$$
$$x_2(t) = -25 e^{-3t/25} + 25 e^{-t/25}.$$

6. Let x_1, x_2, and x_3 be the amounts of salt in tanks A, B, and C, respectively, so that
$$x_1' = \frac{1}{100} x_2 \cdot 2 - \frac{1}{100} x_1 \cdot 6 = \frac{1}{50} x_2 - \frac{3}{50} x_1$$
$$x_2' = \frac{1}{100} x_1 \cdot 6 + \frac{1}{100} x_3 - \frac{1}{100} x_2 \cdot 2 - \frac{1}{100} x_2 \cdot 5 = \frac{3}{50} x_1 - \frac{7}{100} x_2 + \frac{1}{100} x_3$$
$$x_3' = \frac{1}{100} x_2 \cdot 5 - \frac{1}{100} x_3 - \frac{1}{100} x_3 \cdot 4 = \frac{1}{20} x_2 - \frac{1}{20} x_3.$$

9. From the graph we see that the populations are first equal at about $t = 5.6$. The approximate periods of x and y are both 45.

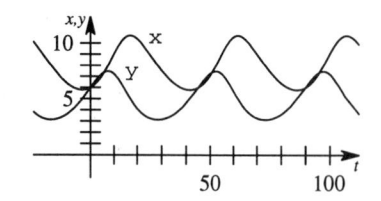

Exercises 4.5

3. From the characteristic equation $\lambda^2 + 8\lambda + 12 = (\lambda+6)(\lambda+2) = 0$ we obtain the eigenvalues $\lambda_1 = -6$, $\lambda_2 = -2$. $(0,0)$ is an asymptotically stable node.

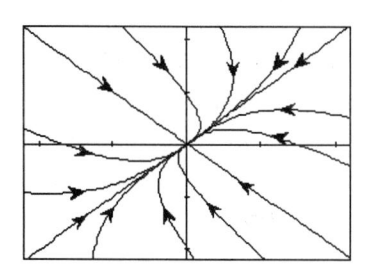

6. From the characteristic equation $\lambda^2 + 4\lambda + 6 = 0$ we obtain the eigenvalues $\lambda_1 = -2 + \sqrt{2}\,i$, $\lambda_2 = -2 - \sqrt{2}\,i$. $(0,0)$ is an asymptotically spiral point.

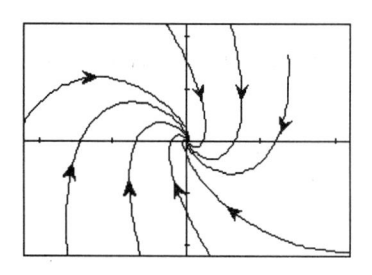

9. Solving
$$2x + 3y = 6$$
$$-x - 2y = -5$$
we find the critical point $(-3, 4)$. Letting $x = X - 3$ and $y = Y + 4$ the system becomes
$$\frac{dX}{dt} = 2X + 3Y$$
$$\frac{dY}{dt} = -X - 2Y.$$
The eigenvalues are $\lambda_1 = -1$, $\lambda_2 = 1$. $(-3, 4)$ is a saddle point.

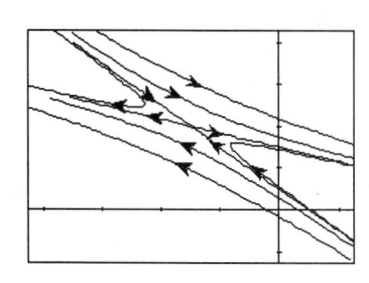

Exercises 4.5

12. Solving
$$3x - 2y = 1$$
$$5x - 3y = 2$$
we find the critical point $(1,1)$. Letting $x = X + 1$ and $y = Y + 1$ the system becomes
$$\frac{dX}{dt} = 3X - 2Y$$
$$\frac{dY}{dt} = 5X - 3Y.$$
The eigenvalues are $\lambda_1 = i$, $\lambda_2 = -i$. $(1,1)$ is a center.

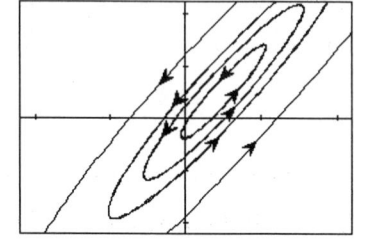

15. The characteristic equation is $\lambda^2 - (1+a)\lambda + a + 4 = 0$ and the eigenvalues are
$\lambda = [1+a \pm \sqrt{(1+a)^2 - 4a - 16}]/2$. Now $(1+a)^2 - 4a - 16 = a^2 - 2a - 15 = (a-5)(a+3) < 0$ for $-3 < a < 5$. Thus, $(0,0)$ is a spiral point for a in the interval $(-3, 5)$. For a to be an asymptotically stable spiral point we need $1 + a < 0$ or $a < -1$. Therefore, $(0,0)$ will be an asymptotically stable spiral point for $-3 < a < -1$.

18. Using $f_x(x,y) = -\cos(x-y)$, $f_y(x,y) = \cos(x-y)$, $g_x(x,y) = e^{x-2y}$, and $g_y(x,y) = -2e^{x-2y} + 5\cos 5y$, we find $a = f_x(0,0) = -1$, $b = f_y(0,0) = 1$, $c = g_x(0,0) = 1$, and $d = g_y(0,0) = 3$. A linearization of the system is
$$\frac{dx}{dt} = -x + y$$
$$\frac{dy}{dt} = x + 3y,$$
and the eigenvalues are $\lambda_1 = 1 - \sqrt{5}$ and $\lambda_2 = 1 + \sqrt{5}$. Thus, $(0,0)$ is a saddle point.

21. Using $f_x(x,y) = \cos x$, $f_y(x,y) = 3\cos y$, $g_x(x,y) = 2\cos x - 2x \sin x$, and $g_y(x,y) = 2y - 2$, we find $a = f_x(0,0) = 1$, $b = f_y(0,0) = 3$, $c = g_x(0,0) = 2$, and $d = g_y(0,0) = -2$. A linearization of the system is
$$\frac{dx}{dt} = x + 3y$$
$$\frac{dy}{dt} = 2x - 2y,$$
and the eigenvalues are $\lambda_1 = -\frac{1}{2} - \frac{1}{2}\sqrt{33}$ and $\lambda_2 = -\frac{1}{2} + \frac{1}{2}\sqrt{33}$. Thus, $(0,0)$ is a saddle point.

24. Using $f_x(x,y) = -\cos x - 1$, $f_y(x,y) = 3$, $g_x(x,y) = -1$, and $g_y(x,y) = 2ye^y + 2e^y$, we find $a = f_x(0,0) = -2$, $b = f_y(0,0) = 3$, $c = g_x(0,0) = -1$, and $d = g_y(0,0) = 2$. A linearization of the

Exercises 4.5

system is

$$\frac{dx}{dt} = -2x + 3y$$

$$\frac{dy}{dt} = -x + 2y,$$

and the eigenvalues are $\lambda_1 = -1$ and $\lambda_2 = 1$. Thus, $(0,0)$ is a saddle point.

27. The critical points are $(1,1)$, $(2,1)$ and $(3/2, 3/2)$. For $(1,1)$ we let $x = X+1$ and $y = Y+1$. Then the system becomes

$$\frac{dX}{dt} = -X^2 + X - XY$$

$$\frac{dY}{dt} = 2XY - Y.$$

Using $f_X(X,Y) = -2X - Y + 1$, $f_Y(X,Y) = -X$, $g_X(X,Y) = 2Y$, and $g_Y(X,Y) = 2X - 1$, we find $a = f_X(0,0) = 1$, $b = f_Y(0,0) = 0$, $c = g_X(0,0) = 0$, and $d = g_Y(0,0) = -1$. A linearization of the system is

$$\frac{dX}{dt} = X$$

$$\frac{dY}{dt} = -Y,$$

and the eigenvalues are $\lambda_1 = -1$ and $\lambda_2 = 1$. Thus $(1,1)$ is a saddle point.

For $(2,1)$ we let $x = X + 2$ and $y = Y + 1$. Then the system becomes

$$\frac{dX}{dt} = -(X+1)(X+Y)$$

$$\frac{dY}{dt} = (2X+1)Y.$$

Using $f_X(X,Y) = -2X - Y - 1$, $f_Y(X,Y) = -X - 1$, $g_X(X,Y) = 2Y$, and $g_Y(X,Y) = 2X + 1$, we find $a = f_X(0,0) = -1$, $b = f_Y(0,0) = -1$, $c = g_X(0,0) = 0$, and $d = g_Y(0,0) = 1$. A linearization of the system is

$$\frac{dX}{dt} = -X - Y$$

$$\frac{dY}{dt} = Y,$$

and the eigenvalues are $\lambda_1 = -1$ and $\lambda_2 = 1$. Thus, $(2,1)$ is a saddle point.

For $(3/2, 3/2)$ we let $x = X + 3/2$ and $y = Y + 3/2$. Then the system becomes

$$\frac{dX}{dt} = \left(X + \frac{1}{2}\right)(-X - Y)$$

$$\frac{dY}{dt} = X + 2XY.$$

Exercises 4.5

Using $f_X(X,Y) = -2X - Y - 1/2$, $f_Y(X,Y) = -X - 1/2$, $g_X(X,Y) = 2Y+1$, and $g_Y(X,Y) = 2X$, we find $a = f_X(0,0) = -1/2$, $b = f_Y(0,0) = -1/2$, $c = g_X(0,0) = 1$, and $d = g_Y(0,0) = 0$. A linearization of the system is

$$\frac{dX}{dt} = -\frac{1}{2}X - \frac{1}{2}Y$$

$$\frac{dY}{dt} = X,$$

and the eigenvalues are $\lambda_1 = -\frac{1}{4} - \frac{\sqrt{7}}{4}i$ and $\lambda_2 = -\frac{1}{4} + \frac{\sqrt{7}}{4}i$. Thus, $(3/2, 3/2)$ is an asymptotically stable spiral point.

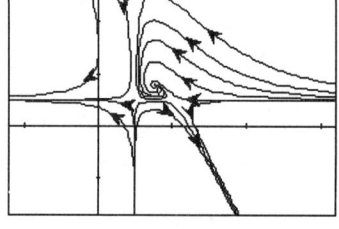

30. The critical points are $(-1, 3)$ and $(-1, -3)$. From the phase portrait we see that $(-1, 3)$ is an unstable spiral point and $(-1, -3)$ is a saddle point.

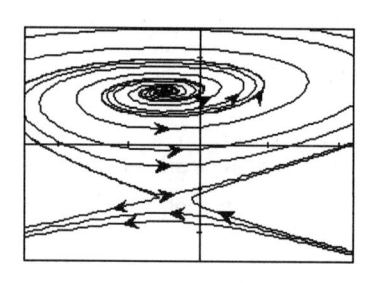

33. (a) The linearized system in each case is

$$\frac{dx}{dt} = y$$

$$\frac{dy}{dt} = -x.$$

(b) The characteristic equation is $\lambda^2 + 1 = 0$ and the eigenvalues are $\lambda = \pm i$. The critical point $(0,0)$ is a center.

(c) We see from the phase portraits that $(0,0)$ is a center for the first system and an asymptotically stable spiral point for the second system.

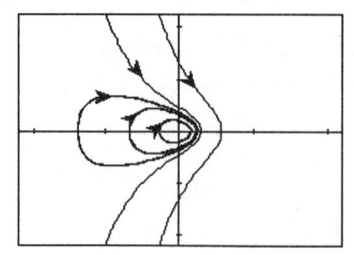

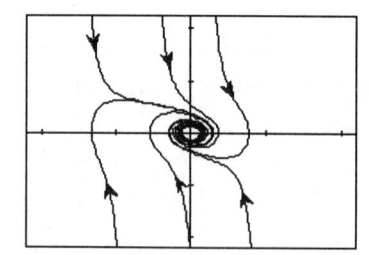

Exercises 4.5

36. The equivalent system is

$$\frac{dx}{dt} = y$$
$$\frac{dy}{dt} = \mu(1 - x^2)y - x,$$

from which we easily see that $(0,0)$ is the only critical point. Linearizing we obtain

$$\frac{dx}{dt} = y$$
$$\frac{dy}{dt} = -x + \mu y.$$

The eigenvalues for this system are $\lambda = \frac{1}{2}(\mu \pm \sqrt{\mu^2 - 4})$. For $\mu^2 - 4 < 0$ or $0 < \mu < 2$, λ is complex with a positive real part $\mu/2$, and $(0,0)$ is an unstable spiral point. For $\mu^2 - 4 > 0$ or $\mu > 2$, both eigenvalues are real and positive. In this case $(0,0)$ is an unstable node.

39. (a) The system is

$$\frac{dx}{dt} = x^2 - y^2$$
$$\frac{dy}{dt} = 2xy.$$

Dividing, we obtain the homogeneous equation $2xy\, dx = (x^2 - y^2)\, dy$. Using the technique in Section 2.2 we obtain the solution $x^2 + y^2 = cy$. Since the trajectory passes through (x_0, y_0) we have $x_0^2 + y_0^2 = cy_0$ or $c = (x_0^2 + y_0^2)/y_0$, $y \neq 0$.

(b) Writing the trajectory in the form $x^2 + (y - c/2)^2 = (c/2)^2$ we see that it is a circle of radius $c/2$ centered on the y-axis and passing through $(0,0)$. Fix $\epsilon > 0$ and let C_ϵ be the circle of radius ϵ centered at $(0,0)$. Now let $T_{2\epsilon}$ be the trajectory passing through the point $(0, 2\epsilon)$. This trajectory passes through points outside C_ϵ [for example, $(0, 2\epsilon)$], but since it is a circle centered on the y-axis and passing through $(0,0)$, it contains points (x_0, y_0) lying inside any circle C_δ of radius δ centered at $(0,0)$. Thus, $(0,0)$ is an unstable critical point.

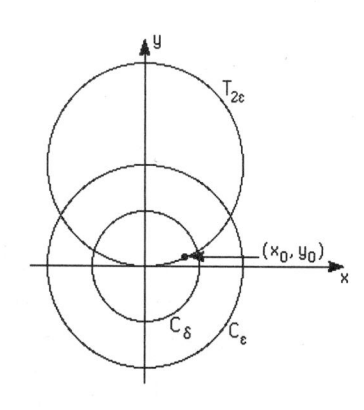

55

Exercises 4.5

42. (a) Critical points are $(0,0)$, $(0,10)$, $(5,0)$, and $(2,4)$. We see from the phase portrait that $(0,0)$ is an unstable node, $(0,10)$ and $(5,0)$ are stable nodes and $(2,4)$ is a saddle point. We thus see that the species will not become extinct simultaneously [$(0,0)$ is unstable], but individually they may become extinct since $x \to 0$ as $y \to 10$ and $y \to 0$ as $x \to 5$. Finally, the species cannot both survive since $(2,4)$ is unstable.

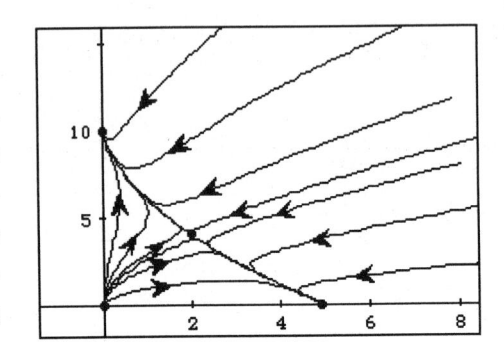

(b) Critical points are $(0,0)$, $(0,17)$, $(10,0)$, and $(6,8)$. We see from the phase portrait that $(0,0)$ is an unstable node, $(0,17)$ and $(10,0)$ are unstable saddle points, and $(6,8)$ is a stable node. We thus see that neither species can become extinct and both populations approach equilibrium since $x \to 6$ and $y \to 8$ as $t \to \infty$.

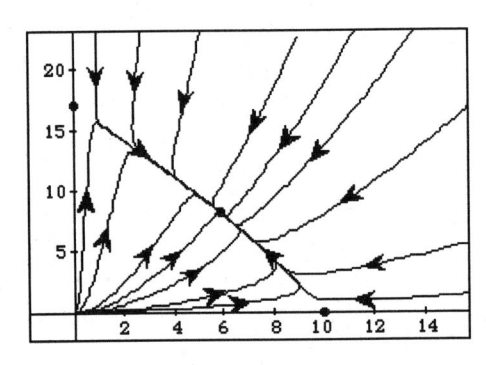

5 The Laplace Transform

Exercises 5.1

3. $\mathcal{L}\{f(t)\} = \int_0^1 te^{-st}dt + \int_1^\infty e^{-st}dt = \left[-\frac{1}{s}te^{-st} - \frac{1}{s^2}e^{-st}\right]_0^1 - \frac{1}{s}e^{-st}\Big|_1^\infty$

$= \left(-\frac{1}{s}e^{-s} - \frac{1}{s^2}e^{-s}\right) - \left(0 - \frac{1}{s^2}\right) - \frac{1}{s}(0 - e^{-s}) = \frac{1}{s^2}(1 - e^{-s}), \quad s > 0$

6. $\mathcal{L}\{f(t)\} = \int_{\pi/2}^\infty (\cos t)e^{-st}dt = \left[-\frac{s}{s^2+1}e^{-st}\cos t + \frac{1}{s^2+1}e^{-st}\sin t\right]_{\pi/2}^\infty$

$= 0 - \left(0 + \frac{1}{s^2+1}e^{-\pi s/2}\right) = -\frac{1}{s^2+1}e^{-\pi s/2}, \quad s > 0$

9. $f(t) = \begin{cases} 1-t, & 0 < t < 1 \\ 0, & t > 0 \end{cases}$

$\mathcal{L}\{f(t)\} = \int_0^1 (1-t)e^{-st}dt = \left[-\frac{1}{s}(1-t)e^{-st} + \frac{1}{s^2}e^{-st}\right]_0^1 = \frac{1}{s^2}e^{-s} + \frac{1}{s} - \frac{1}{s^2}, \quad s > 0$

12. $\mathcal{L}\{f(t)\} = \int_0^\infty e^{-2t-5}e^{-st}dt = e^{-5}\int_0^\infty e^{-(s+2)t}dt = -\frac{e^{-5}}{s+2}e^{-(s+2)t}\Big|_0^\infty = \frac{e^{-5}}{s+2}, \quad s > -2$

15. $\mathcal{L}\{f(t)\} = \int_0^\infty (\sin kt)e^{-st}dt = \left[-\frac{s}{s^2+k^2}(\sin kt)e^{-st} - \frac{k}{s^2+k^2}(\cos kt)e^{-st}\right]_0^\infty = \frac{k}{s^2+k^2}, \quad s > 0$

18. $\mathcal{L}\{f(t)\} = \int_0^\infty t(\sin t)e^{-st}dt$

$= \left[\left(-\frac{t}{s^2+1} - \frac{2s}{(s^2+1)^2}\right)(\cos t)e^{-st} - \left(\frac{st}{s^2+1} + \frac{s^2-1}{(s^2+1)^2}\right)(\sin t)e^{-st}\right]_0^\infty$

$= \frac{2s}{(s^2+1)^2}, \quad s > 0$

21. $\mathcal{L}\{1 + e^{4t}\} = \frac{1}{s} + \frac{1}{s-4}$

24. $\mathcal{L}\{e^{2t} - 2 + e^{-2t}\} = \frac{1}{s-2} - \frac{2}{s} + \frac{1}{s+2}$

27. $\mathcal{L}\{\sinh kt\} = \mathcal{L}\left\{\frac{1}{2}(e^{kt} - e^{-kt})\right\} = \frac{1}{2}\left(\frac{1}{s-k} - \frac{1}{s+k}\right) = \frac{k}{s^2 - k^2}$

30. $\mathcal{L}\{\sin^2 4t\} = \mathcal{L}\left\{\frac{1}{2} - \frac{1}{2}\cos 8t\right\} = \frac{1}{2s} + \frac{1}{2}\frac{s}{s^2+64}$

Exercises 5.1

33. Let $f(t) = 1$ and $g(t) = \begin{cases} 1, & t \geq 0, \ t \neq 1 \\ 0, & t = 1 \end{cases}$. Then $\mathscr{L}\{f(t)\} = \mathscr{L}\{g(t)\} = 1$, but $f(t) \neq g(t)$.

36. The relation will be valid when s is greater than the maximum of c_1 and c_2.

39. $\mathscr{L}^{-1}\left\{\dfrac{1}{s^3}\right\} = \dfrac{1}{2}\mathscr{L}^{-1}\left\{\dfrac{2}{s^3}\right\} = \dfrac{1}{2}t^2$

42. $\mathscr{L}^{-1}\left\{\left(\dfrac{2}{s} - \dfrac{1}{s^3}\right)^2\right\} = \mathscr{L}^{-1}\left\{4 \cdot \dfrac{1}{s^2} - \dfrac{4}{6} \cdot \dfrac{3!}{s^4} + \dfrac{1}{120} \cdot \dfrac{5!}{s^6}\right\} = 4t - \dfrac{2}{3}t^3 + \dfrac{1}{120}t^5$

45. $\mathscr{L}^{-1}\left\{\dfrac{1}{s^2} - \dfrac{1}{s} + \dfrac{1}{s-2}\right\} = t - 1 + e^{2t}$

48. $\mathscr{L}^{-1}\left\{\dfrac{1}{5s-2}\right\} = \mathscr{L}^{-1}\left\{\dfrac{1}{5} \cdot \dfrac{1}{s-2/5}\right\} = \dfrac{1}{5}e^{2t/5}$

51. $\mathscr{L}^{-1}\left\{\dfrac{4s}{4s^2+1}\right\} = \mathscr{L}^{-1}\left\{\dfrac{s}{s^2+1/4}\right\} = \cos\dfrac{1}{2}t$

54. $\mathscr{L}^{-1}\left\{\dfrac{10s}{s^2-25}\right\} = 10\cosh 5t$

57. $\mathscr{L}^{-1}\left\{\dfrac{1}{s^2+3s}\right\} = \mathscr{L}^{-1}\left\{\dfrac{1}{3} \cdot \dfrac{1}{s} - \dfrac{1}{3} \cdot \dfrac{1}{s+3}\right\} = \dfrac{1}{3} - \dfrac{1}{3}e^{-3t}$

60. $\mathscr{L}^{-1}\left\{\dfrac{1}{s^2+s-20}\right\} = \mathscr{L}^{-1}\left\{\dfrac{1}{9} \cdot \dfrac{1}{s-4} - \dfrac{1}{9} \cdot \dfrac{1}{s+5}\right\} = \dfrac{1}{9}e^{4t} - \dfrac{1}{9}e^{-5t}$

63. $\mathscr{L}^{-1}\left\{\dfrac{s}{(s-2)(s-3)(s-6)}\right\} = \mathscr{L}^{-1}\left\{\dfrac{1}{2} \cdot \dfrac{1}{s-2} - \dfrac{1}{s-3} + \dfrac{1}{2} \cdot \dfrac{1}{s-6}\right\} = \dfrac{1}{2}e^{2t} - e^{3t} + \dfrac{1}{2}e^{6t}$

66. $\mathscr{L}^{-1}\left\{\dfrac{s+1}{(s^2-4s)(s+5)}\right\} = \mathscr{L}^{-1}\left\{-\dfrac{1}{20} \cdot \dfrac{1}{s} + \dfrac{5}{36} \cdot \dfrac{1}{s-4} - \dfrac{4}{45} \cdot \dfrac{1}{s+5}\right\} = -\dfrac{1}{20} + \dfrac{5}{36}e^{4t} - \dfrac{4}{45}e^{-5t}$

69. $\mathscr{L}^{-1}\left\{\dfrac{s}{(s^2+4)(s+2)}\right\} = \mathscr{L}^{-1}\left\{\dfrac{1}{4} \cdot \dfrac{s}{s^2+4} + \dfrac{1}{4} \cdot \dfrac{2}{s^2+4} - \dfrac{1}{4} \cdot \dfrac{1}{s+2}\right\} = \dfrac{1}{4}\cos 2t + \dfrac{1}{4}\sin 2t - \dfrac{1}{4}e^{-2t}$

72. The Laplace transform of the differential equation is

$$s\mathscr{L}\{y\} - y(0) + 2\mathscr{L}\{y\} = \dfrac{1}{s^2}.$$

Solving for $\mathscr{L}\{y\}$ we obtain

$$\mathscr{L}\{y\} = \dfrac{1-s^2}{s^2(s+2)} = -\dfrac{1}{4}\dfrac{1}{s} + \dfrac{1}{2}\dfrac{1}{s^2} - \dfrac{3}{4}\dfrac{1}{s+2}.$$

Thus

$$y = -\dfrac{1}{4} + \dfrac{1}{2}t - \dfrac{3}{4}e^{-2t}.$$

75. The Laplace transform of the differential equation is

$$s^2 \mathcal{L}\{y\} - sy(0) - y'(0) + 5[s\mathcal{L}\{y\} - y(0)] + 4\mathcal{L}\{y\} = 0.$$

Solving for $\mathcal{L}\{y\}$ we obtain

$$\mathcal{L}\{y\} = \frac{s+5}{s^2+5s+4} = \frac{4}{3}\frac{1}{s+1} - \frac{1}{3}\frac{1}{s+4}.$$

Thus

$$y = \frac{4}{3}e^{-t} - \frac{1}{3}e^{-4t}.$$

78. The Laplace transform of the differential equation is

$$s^3\mathcal{L}\{y\} - s^2(0) - sy'(0) - y''(0) + 2[s^2\mathcal{L}\{y\} - sy(0) - y'(0)] - [s\mathcal{L}\{y\} - y(0)] - 2\mathcal{L}\{y\} = \frac{3}{s^2+9}.$$

Solving for $\mathcal{L}\{y\}$ we obtain

$$\mathcal{L}\{y\} = \frac{s^2+12}{(s-1)(s+1)(s+2)(s^2+9)}$$

$$= \frac{13}{60}\frac{1}{s-1} - \frac{13}{20}\frac{1}{s+1} + \frac{16}{39}\frac{1}{s+2} + \frac{3}{130}\frac{s}{s^2+9} - \frac{1}{65}\frac{3}{s^2+9}.$$

Thus

$$y = \frac{13}{60}e^t - \frac{13}{20}e^{-t} + \frac{16}{39}e^{-2t} + \frac{3}{130}\cos 3t - \frac{1}{65}\sin 3t.$$

Exercises 5.2

3. $\mathcal{L}\{t^3 e^{-2t}\} = \dfrac{3!}{(s+2)^4}$

6. $\mathcal{L}\{e^{-2t}\cos 4t\} = \dfrac{s+2}{(s+2)^2+16}$

9. $\mathcal{L}\{t(e^t+e^{2t})^2\} = \mathcal{L}\{te^{2t}+2te^{3t}+te^{4t}\} = \dfrac{1}{(s-2)^2} + \dfrac{2}{(s-3)^2} + \dfrac{1}{(s-4)^2}$

12. $\mathcal{L}\{e^t\cos^2 3t\} = \mathcal{L}\left\{\dfrac{1}{2}e^t + \dfrac{1}{2}e^t\cos 6t\right\} = \dfrac{1}{2}\dfrac{1}{s-1} + \dfrac{1}{2}\dfrac{s-1}{(s-1)^2+36}$

15. $\mathcal{L}^{-1}\left\{\dfrac{1}{s^2-6s+10}\right\} = \mathcal{L}^{-1}\left\{\dfrac{1}{(s-3)^2+1^2}\right\} = e^{3t}\sin t$

18. $\mathcal{L}^{-1}\left\{\dfrac{2s+5}{s^2+6s+34}\right\} = \mathcal{L}^{-1}\left\{2\dfrac{(s+3)}{(s+3)^2+5^2} - \dfrac{1}{5}\dfrac{5}{(s+3)^2+5^2}\right\} = 2e^{-3t}\cos 5t - \dfrac{1}{5}e^{-3t}\sin 5t$

Exercises 5.2

21. $\mathcal{L}^{-1}\left\{\dfrac{2s-1}{s^2(s+1)^3}\right\} = \mathcal{L}^{-1}\left\{\dfrac{5}{s} - \dfrac{1}{s^2} - \dfrac{5}{s+1} - \dfrac{4}{(s+1)^2} - \dfrac{3}{2}\dfrac{2}{(s+1)^3}\right\} = 5 - t - 5e^{-t} - 4te^{-t} - \dfrac{3}{2}t^2 e^{-t}$

24. $\mathcal{L}\{e^{2-t}\,\mathcal{U}(t-2)\} = \mathcal{L}\{e^{-(t-2)}\,\mathcal{U}(t-2)\} = \dfrac{e^{-2s}}{s+1}$

27. $\mathcal{L}\{\cos 2t\,\mathcal{U}(t-\pi)\} = \mathcal{L}\{\cos 2(t-\pi)\,\mathcal{U}(t-\pi)\} = \dfrac{se^{-\pi s}}{s^2+4}$

30. $\mathcal{L}\{te^{t-5}\,\mathcal{U}(t-5)\} = \mathcal{L}\{(t-5)e^{t-5}\,\mathcal{U}(t-5) + 5e^{t-5}\,\mathcal{U}(t-5)\} = \dfrac{e^{-5s}}{(s-1)^2} + \dfrac{5e^{-5s}}{s-1}$

33. $\mathcal{L}^{-1}\left\{\dfrac{e^{-\pi s}}{s^2+1}\right\} = \sin(t-\pi)\,\mathcal{U}(t-\pi)$

36. $\mathcal{L}^{-1}\left\{\dfrac{e^{-2s}}{s^2(s-1)}\right\} = \mathcal{L}^{-1}\left\{-\dfrac{e^{-2s}}{s} - \dfrac{e^{-2s}}{s^2} + \dfrac{e^{-2s}}{s-1}\right\} = -\mathcal{U}(t-2) - (t-2)\mathcal{U}(t-2) + e^{t-2}\mathcal{U}(t-2)$

39. (f)

42. (d)

45. $\mathcal{L}\{t^2\,\mathcal{U}(t-1)\} = \mathcal{L}\{[(t-1)^2 + 2t - 1]\,\mathcal{U}(t-1)\} = \mathcal{L}\{[(t-1)^2 + 2(t-1) + 1]\,\mathcal{U}(t-1)\}$

$= \left(\dfrac{2}{s^3} + \dfrac{2}{s^2} + \dfrac{1}{s}\right)e^{-s}$

48. $\mathcal{L}\{\sin t - \sin t\,\mathcal{U}(t-2\pi)\} = \mathcal{L}\{\sin t - \sin(t-2\pi)\,\mathcal{U}(t-2\pi)\} = \dfrac{1}{s^2+1} - \dfrac{e^{-2\pi s}}{s^2+1}$

51. The Laplace transform of the differential equation is

$$s^2\mathcal{L}\{y\} - sy(0) - y'(0) - 6[s\mathcal{L}\{y\} - y(0)] + 13\mathcal{L}\{y\} = 0.$$

Solving for $\mathcal{L}\{y\}$ we obtain

$$\mathcal{L}\{y\} = -\dfrac{3}{s^2 - 6s + 13} = -\dfrac{3}{2}\dfrac{2}{(s-3)^2 + 2^2}.$$

Thus $y = -\dfrac{3}{2}e^{3t}\sin 2t.$

54. The Laplace transform of the differential equation is

$$s^2\mathcal{L}\{y\} - sy(0) - y'(0) - 2[s\mathcal{L}\{y\} - y(0)] + 5\mathcal{L}\{y\} = \dfrac{1}{s} + \dfrac{1}{s^2}.$$

Solving for $\mathcal{L}\{y\}$ we obtain

$$\mathcal{L}\{y\} = \dfrac{4s^2 + s + 1}{s^2(s^2 - 2s + 5)} = \dfrac{7}{25}\dfrac{1}{s} + \dfrac{1}{5}\dfrac{1}{s^2} + \dfrac{-7s/25 + 109/25}{s^2 - 2s + 5}$$

$$= \dfrac{7}{25}\dfrac{1}{s} + \dfrac{1}{5}\dfrac{1}{s^2} - \dfrac{7}{25}\dfrac{s-1}{(s-1)^2 + 2^2} + \dfrac{51}{25}\dfrac{2}{(s-1)^2 + 2^2}.$$

Thus
$$y = \frac{7}{25} + \frac{1}{5}t - \frac{7}{25}e^t \cos 2t + \frac{51}{25}e^t \sin 2t.$$

57. The Laplace transform of the differential equation is
$$s\mathcal{L}\{y\} - y(0) + \mathcal{L}\{y\} = \frac{5}{s}e^{-s}.$$

Solving for $\mathcal{L}\{y\}$ we obtain
$$\mathcal{L}\{y\} = \frac{5e^{-s}}{s(s+1)} = 5e^{-s}\left[\frac{1}{s} - \frac{1}{s+1}\right].$$

Thus
$$y = 5\,\mathcal{U}(t-1) - 5e^{-(t-1)}\mathcal{U}(t-1).$$

60. The Laplace transform of the differential equation is
$$s^2\mathcal{L}\{y\} - sy(0) - y'(0) + 4\mathcal{L}\{y\} = \frac{1}{s} - \frac{e^{-s}}{s}.$$

Solving for $\mathcal{L}\{y\}$ we obtain
$$\mathcal{L}\{y\} = \frac{1-s}{s(s^2+4)} - e^{-s}\frac{1}{s(s^2+4)} = \frac{1}{4}\frac{1}{s} - \frac{1}{4}\frac{s}{s^2+4} - \frac{1}{2}\frac{2}{s^2+4} - e^{-s}\left[\frac{1}{4}\frac{1}{s} - \frac{1}{4}\frac{s}{s^2+4}\right].$$

Thus
$$y = \frac{1}{4} - \frac{1}{4}\cos 2t - \frac{1}{2}\sin 2t - \left[\frac{1}{4} - \frac{1}{4}\cos 2(t-1)\right]\mathcal{U}(t-1).$$

63. The Laplace transform of the differential equation is
$$s^2\mathcal{L}\{y\} - sy(0) - y'(0) + \mathcal{L}\{y\} = \frac{e^{-\pi s}}{s} - \frac{e^{-2\pi s}}{s}.$$

Solving for $\mathcal{L}\{y\}$ we obtain
$$\mathcal{L}\{y\} = e^{-\pi s}\left[\frac{1}{s} - \frac{s}{s^2+1}\right] - e^{-2\pi s}\left[\frac{1}{s} - \frac{s}{s^2+1}\right] + \frac{1}{s^2+1}.$$

Thus
$$y = [1 - \cos(t-\pi)]\mathcal{U}(t-\pi) - [1 - \cos(t-2\pi)]\mathcal{U}(t-2\pi) + \sin t.$$

66. Taking the Laplace transform of both sides of the differential equation and letting $c = y'(0)$ we obtain
$$\mathcal{L}\{y''\} - \mathcal{L}\{9y'\} + \mathcal{L}\{20y\} = \mathcal{L}\{1\}$$
$$s^2\mathcal{L}\{y\} - sy(0) - y'(0) - 9s\mathcal{L}\{y\} + 9y(0) + 20\mathcal{L}\{y\} = \frac{1}{s}$$
$$s^2\mathcal{L}\{y\} - c - 9s\mathcal{L}\{y\} + 20\mathcal{L}\{y\} = \frac{1}{s}$$
$$(s^2 - 9s + 20)\mathcal{L}\{y\} = \frac{1}{s} + c.$$

Exercises 5.2

Solving for $\mathcal{L}\{y\}$ we obtain

$$\mathcal{L}\{y\} = \frac{1}{s(s^2 - 9s + 20)} + \frac{c}{s^2 - 9s + 20}$$

$$= \frac{1}{s(s-4)(s-5)} + \frac{c}{(s-4)(s-5)}$$

$$= \frac{1/20}{s} - \frac{1/4}{s-4} + \frac{1/5}{s-5} - \frac{c}{s-4} + \frac{c}{s-5}.$$

Therefore

$$y(t) = \frac{1}{20}\mathcal{L}^{-1}\left\{\frac{1}{s}\right\} - \frac{1}{4}\mathcal{L}^{-1}\left\{\frac{1}{s-4}\right\} + \frac{1}{5}\mathcal{L}^{-1}\left\{\frac{1}{s-5}\right\} - c\mathcal{L}^{-1}\left\{\frac{1}{s-4}\right\} + c\mathcal{L}^{-1}\left\{\frac{1}{s-5}\right\}$$

$$= \frac{1}{20} - \frac{1}{4}e^{4t} + \frac{1}{5}e^{5t} - c\left(e^{4t} - e^{5t}\right).$$

To find c we compute

$$y'(t) = -e^{4t} + e^{5t} - c\left(4e^{4t} - 5e^{5t}\right)$$

and let $y'(1) = 0$. Then

$$0 = -e^4 + e^5 - c\left(4e^4 - 5e^5\right)$$

and

$$c = \frac{e^5 - e^4}{4e^4 - 5e^5} = \frac{e - 1}{4 - 5e}.$$

Thus

$$y(t) = \frac{1}{20} - \frac{1}{4}e^{4t} + \frac{1}{5}e^{5t} - \frac{e-1}{4-5e}\left(e^{4t} - e^{5t}\right)$$

$$= \frac{1}{20} + \frac{e}{4(4-5e)}e^{4t} - \frac{1}{5(4-5e)}e^{5t}.$$

69. The differential equation is

$$2.5\frac{dq}{dt} + 12.5q = 5\,\mathcal{U}(t - 3).$$

The Laplace transform of this equation is

$$s\mathcal{L}\{q\} + 5\mathcal{L}\{q\} = \frac{2}{s}e^{-3s}.$$

Solving for $\mathcal{L}\{q\}$ we obtain

$$\mathcal{L}\{q\} = \frac{2}{s(s+5)}e^{-3s} = \left(\frac{2}{5}\cdot\frac{1}{s} - \frac{2}{5}\cdot\frac{1}{s+5}\right)e^{-3s}.$$

Thus

$$q(t) = \frac{2}{5}\mathcal{U}(t-3) - \frac{2}{5}e^{-5(t-3)}\mathcal{U}(t-3).$$

62

Exercises 5.2

72. The differential equation is

$$\frac{d^2q}{dt^2} + 20\frac{dq}{dt} + 200q = 150, \quad q(0) = q'(0) = 0.$$

The Laplace transform of this equation is

$$s^2 \mathscr{L}\{q\} + 20s\mathscr{L}\{q\} + 200\mathscr{L}\{q\} = \frac{150}{s}.$$

Solving for $\mathscr{L}\{q\}$ we obtain

$$\mathscr{L}\{q\} = \frac{150}{s(s^2 + 20s + 200)} = \frac{3}{4}\frac{1}{s} - \frac{3}{4}\frac{s+10}{(s+10)^2 + 10^2} - \frac{3}{4}\frac{10}{(s+10)^2 + 10^2}.$$

Thus

$$q(t) = \frac{3}{4} - \frac{3}{4}e^{-10t}\cos 10t - \frac{3}{4}e^{-10t}\sin 10t$$

and

$$i(t) = q'(t) = 15e^{-10t}\sin 10t.$$

If $E(t) = 150 - 150\,\mathscr{U}(t-2)$, then

$$\mathscr{L}\{q\} = \frac{150}{s(s^2+20s+200)}\left(1 - e^{-2s}\right)$$

$$q(t) = \frac{3}{4} - \frac{3}{4}e^{-10t}\cos 10t - \frac{3}{4}e^{-10t}\sin 10t - \left[\frac{3}{4} - \frac{3}{4}e^{-10(t-2)}\cos 10(t-2)\right.$$

$$\left. - \frac{3}{4}e^{-10(t-2)}\sin 10(t-2)\right]\mathscr{U}(t-2).$$

75. Recall from Chapter 3 that $mx'' = -kx - \beta x'$. Now $m = W/g = 4/32 = \frac{1}{8}$ slug, and $4 = 2k$ so that $k = 2$ lb./ft. Thus, the differential equation is $x'' + 7x' + 16x = 0$. The initial conditions are $x(0) = -3/2$ and $x'(0) = 0$. The Laplace transform of the differential equation is

$$s^2\mathscr{L}\{x\} + \frac{3}{2}s + 7s\mathscr{L}\{x\} + \frac{21}{2} + 16\mathscr{L}\{x\} = 0.$$

Solving for $\mathscr{L}\{x\}$ we obtain

$$\mathscr{L}\{x\} = \frac{-3s/2 - 21/2}{s^2 + 7s + 16} = -\frac{3}{2}\frac{s+7/2}{(s+7/2)^2 + (\sqrt{15}/2)^2} - \frac{7\sqrt{15}}{10}\frac{\sqrt{15}/2}{(s+7/2)^2 + (\sqrt{15}/2)^2}.$$

Thus

$$x = -\frac{3}{2}e^{-7t/2}\cos\frac{\sqrt{15}}{2}t - \frac{7\sqrt{15}}{10}e^{-7t/2}\sin\frac{\sqrt{15}}{2}t.$$

78. The differential equation is

$$EI\frac{d^4y}{dx^4} = w_0\left[2 - \frac{2x}{L} - \left(1 - \frac{2x}{L}\right)\mathscr{U}\left(x - \frac{L}{2}\right)\right] = \frac{2w_0}{L}\left[L - x + \left(x - \frac{L}{2}\right)\mathscr{U}\left(x - \frac{L}{2}\right)\right].$$

63

Exercises 5.2

Taking the Laplace transform of both sides and using $y(0) = y'(0) = 0$ we obtain

$$s^4 \mathcal{L}\{y\} - sy''(0) - y'''(0) = \frac{2w_0}{EIL}\left[\frac{L}{s} - \frac{1}{s^2} + \frac{1}{s^2}e^{-Ls/2}\right].$$

Letting $y''(0) = c_1$ and $y'''(0) = c_2$ we have

$$\mathcal{L}\{y\} = \frac{c_1}{s^3} + \frac{c_2}{s^4} + \frac{2w_0}{EIL}\left[\frac{L}{s^5} - \frac{1}{s^6} + \frac{1}{s^6}e^{-Ls/2}\right]$$

so that

$$y(x) = \frac{1}{2}c_1 x^2 + \frac{1}{6}c_2 x^3 + \frac{2w_0}{EIL}\left[\frac{L}{24}x^4 - \frac{L}{120}x^5 + \frac{1}{120}\left(x - \frac{L}{2}\right)^5 \mathcal{U}\left(x - \frac{L}{2}\right)\right]$$

$$= \frac{1}{2}c_1 x^2 + \frac{1}{6}c_2 x^3 + \frac{w_0}{60EIL}\left[5Lx^4 - Lx^5 + \left(x - \frac{L}{2}\right)^5 \mathcal{U}\left(x - \frac{L}{2}\right)\right].$$

To find c_1 and c_2 we compute

$$y''(x) = c_1 + c_2 x + \frac{w_0}{3EIL}\left[3Lx^2 - Lx^3 + \left(x - \frac{L}{2}\right)^3 \mathcal{U}\left(x - \frac{L}{2}\right)\right]$$

and

$$y''' = c_2 + \frac{w_0}{EIL}\left[2Lx - Lx^2 + \left(x - \frac{L}{2}\right)^2 \mathcal{U}\left(x - \frac{L}{2}\right)\right].$$

Then $y''(L) = y'''(L) = 0$ yields the system

$$c_1 + c_2 L + \frac{25w_0}{24EI}L^2 - \frac{w_0}{3EI}L^3 = 0$$

$$c_2 + \frac{9w_0}{4EI}L - \frac{w_0}{EI}L^2 = 0.$$

Solving for c_1 and c_2 we obtain $c_1 = \frac{w_0 L^3}{3EI} - \frac{19w_0 L^2}{24EI}$ and $c_2 = -\frac{w_0 L}{4EI}$. Thus

$$y(x) = \frac{w_0(8L - 19)L^2}{48EI} - \frac{w_0 L}{24EI}x^3 + \frac{w_0}{12EI}x^4 - \frac{w_0}{60EI}x^5 + \frac{w_0}{60EIL}\left(x - \frac{L}{2}\right)^5 \mathcal{U}\left(x - \frac{L}{2}\right).$$

Exercises 5.3

3. $\mathcal{L}\{t^2 \sinh t\} = \dfrac{d^2}{ds^2}\left(\dfrac{1}{s^2 - 1}\right) = \dfrac{6s^2 + 2}{(s^2 - 1)^3}$

6. $\mathcal{L}\{te^{-3t}\cos 3t\} = -\dfrac{d}{ds}\left(\dfrac{s + 3}{(s+3)^2 + 9}\right) = \dfrac{(s+3)^2 - 9}{[(s+3)^2 + 9]^2}$

9. $\mathcal{L}\left\{\displaystyle\int_0^t e^\tau\, d\tau\right\} = \dfrac{1}{s}\mathcal{L}\{e^t\} = \dfrac{1}{s(s-1)}$

64

Exercises 5.3

12. $\mathscr{L}\left\{\int_0^t \tau \sin \tau \, d\tau\right\} = \frac{1}{s} \mathscr{L}\{t \sin t\} = \frac{1}{s}\left(-\frac{d}{ds}\frac{1}{s^2+1}\right) = -\frac{1}{s}\frac{-2s}{(s^2+1)^2} = \frac{2}{(s^2+1)^2}$

15. $\mathscr{L}\left\{t\int_0^t \sin \tau \, d\tau\right\} = -\frac{d}{ds}\mathscr{L}\left\{\int_0^t \sin \tau \, d\tau\right\} = -\frac{d}{ds}\left(\frac{1}{s}\frac{1}{s^2+1}\right) = \frac{3s^2+1}{s^2(s^2+1)^2}$

18. $\mathscr{L}\{1 * e^{-2t}\} = \dfrac{1}{s(s+2)}$

21. $\mathscr{L}\{e^{-t} * e^t \cos t\} = \dfrac{s-1}{(s+1)[(s-1)^2+1]}$

24. $f(t) = -\dfrac{1}{t}\mathscr{L}^{-1}\left\{\dfrac{d}{ds}\left[\ln\left(s^2+1\right) - \ln\left(s^2+4\right)\right]\right\} = -\dfrac{1}{t}\mathscr{L}^{-1}\left\{\dfrac{2s}{s^2+1} - \dfrac{2s}{s^2+2^2}\right\}$

$= -\dfrac{1}{t}(2\cos t - 2\cos 2t)$

27. $\mathscr{L}^{-1}\left\{\dfrac{1}{s(s+1)}\right\} = 1 * e^{-t} = \int_0^t e^{-(t-\tau)}d\tau = e^{-(t-\tau)}\Big|_0^t = 1 - e^{-t}$

30. $\mathscr{L}^{-1}\left\{\dfrac{1}{(s+1)^2}\right\} = e^{-t} * e^{-t} = \int_0^t e^{-\tau}e^{-(t-\tau)}d\tau = e^{-t}\int_0^t d\tau = te^{-t}$

33. Let $u = t - \tau$ so that $du = d\tau$ and

$$f * g = \int_0^t f(\tau)g(t-\tau)\,d\tau = -\int_t^0 f(t-u)g(u)\,du = g * f.$$

36. The Laplace transform of the differential equation is

$$s^2 \mathscr{L}\{y\} - sy(0) - y'(0) + \mathscr{L}\{y\} = \dfrac{1}{s^2+1}.$$

Letting $y(0) = 1$ and $y'(0) = -1$ and solving for $\mathscr{L}\{y\}$ we obtain

$$\mathscr{L}\{y\} = \dfrac{s^3 - s^2 + s}{(s^2+1)^2} = \dfrac{s}{s^2+1} - \dfrac{1}{s^2+1} + \dfrac{1}{(s^2+1)^2}.$$

Thus $\qquad y = \cos t - \dfrac{1}{2}\sin t - \dfrac{1}{2}t\cos t.$

39. The Laplace transform of the given equation is

$$\mathscr{L}\{f\} + \mathscr{L}\{t\}\mathscr{L}\{f\} = \mathscr{L}\{t\}.$$

Solving for $\mathscr{L}\{f\}$ we obtain $\mathscr{L}\{f\} = \dfrac{1}{s^2+1}$. Thus, $f(t) = \sin t$.

42. The Laplace transform of the given equation is

$$\mathscr{L}\{f\} + 2\mathscr{L}\{\cos t\}\mathscr{L}\{f\} = 4\mathscr{L}\{e^{-t}\} + \mathscr{L}\{\sin t\}.$$

65

Exercises 5.3

Solving for $\mathscr{L}\{f\}$ we obtain

$$\mathscr{L}\{f\} = \frac{4s^2 + s + 5}{(s+1)^3} = \frac{4}{s+1} - \frac{7}{(s+1)^2} + 4\frac{2}{(s+1)^3}.$$

Thus
$$f(t) = 4e^{-t} - 7te^{-t} + 4t^2 e^{-t}.$$

45. The Laplace transform of the given equation is

$$\mathscr{L}\{f\} = \mathscr{L}\{1\} + \mathscr{L}\{t\} + \frac{8}{3}\mathscr{L}\{t^3\}\mathscr{L}\{f\}.$$

Solving for $\mathscr{L}\{f\}$ we obtain

$$\mathscr{L}\{f\} = \frac{s^2(s+1)}{s^4 - 16} = \frac{3}{8}\frac{1}{s-2} + \frac{1}{8}\frac{1}{s+2} + \frac{1}{2}\frac{s}{s^2+4} + \frac{1}{4}\frac{2}{s^2+4}.$$

Thus
$$f(t) = \frac{3}{8}e^{2t} + \frac{1}{8}e^{-2t} + \frac{1}{2}\cos 2t + \frac{1}{4}\sin 2t.$$

48. The Laplace transform of the given equation is

$$s\mathscr{L}\{y\} - y(0) + 6\mathscr{L}\{y\} + 9\mathscr{L}\{1\}\mathscr{L}\{y\} = \mathscr{L}\{1\}.$$

Solving for $\mathscr{L}\{f\}$ we obtain $\mathscr{L}\{y\} = \dfrac{1}{(s+3)^2}$. Thus, $y = te^{-3t}$.

51. The differential equation of the spring/mass system is $mx'' = -kx + f(t)$. Now $m = W/g = 16/32 = 1/2$ slug, and $k = 4.5$, so the differential equation is

$$\frac{1}{2}x'' + 4.5x = 4\sin 3t + 2\cos 3t \quad \text{or} \quad x'' + 9x = 8\sin 3t + 4\cos 3t.$$

The initial conditions are $x(0) = x'(0) = 0$. The Laplace transform of the differential equation is

$$s^2\mathscr{L}\{x\} + 9\mathscr{L}\{x\} = \frac{24}{s^2+9} + \frac{4s}{s^2+9}.$$

Solving for $\mathscr{L}\{x\}$ we obtain

$$\mathscr{L}\{x\} = \frac{4s + 24}{(s^2+9)^2} = \frac{2}{3}\frac{2(3)s}{(s^2+9)^2} + \frac{12}{27}\frac{2(3)^3}{(s^2+9)^2}.$$

Thus
$$x(t) = \frac{2}{3}t\sin 3t + \frac{4}{9}(\sin 3t - 3t\cos 3t) = \frac{2}{3}t\sin 3t + \frac{4}{9}\sin 3t - \frac{4}{3}t\cos 3t.$$

Exercises 5.4

3. $\mathcal{L}\{f(t)\} = \dfrac{1}{1-e^{-2as}} \left[\int_0^a e^{-st} dt - \int_a^{2a} e^{-st} dt \right] = \dfrac{(1-e^{-as})^2}{s(1-e^{-2as})} = \dfrac{1-e^{-as}}{s(1+e^{-as})}$

6. $\mathcal{L}\{f(t)\} = \dfrac{1}{1-e^{-2s}} \left[\int_0^1 te^{-st} dt + \int_1^2 (2-t)e^{-st} dt \right] = \dfrac{1-e^{-s}}{s^2(1-e^{-2s})}$

9. The differential equation is

$$\frac{di}{dt} + \frac{R}{L}i = \frac{1}{L}E(t), \quad i(0) = 0.$$

The Laplace transform of this equation is

$$s\mathcal{L}\{i\} + \frac{R}{L}\mathcal{L}\{i\} = \frac{1}{L}\mathcal{L}\{E(t)\}.$$

From Problem 5 of this section we have

$$\mathcal{L}\{E(t)\} = \frac{1}{s}\left(\frac{1}{s} - \frac{1}{e^s-1}\right) = \frac{1}{s^2} + \frac{1}{s}\frac{1}{1-e^s}.$$

Thus
$$\left(s + \frac{R}{L}\right)\mathcal{L}\{i\} = \frac{1}{L}\frac{1}{s^2} + \frac{1}{L}\frac{1}{s}\frac{1}{1-e^s}$$

and

$$\mathcal{L}\{i\} = \frac{1}{L}\frac{1}{s^2(s+R/L)} + \frac{1}{L}\frac{1}{s(s+R/L)}\frac{1}{1-e^s}$$

$$= \frac{1}{L}\left(\frac{L/R}{s^2} - \frac{L^2/R^2}{s} + \frac{L^2/R^2}{s+R/L}\right) + \frac{1}{L}\left(\frac{L/R}{s} - \frac{L/R}{s+R/L}\right)\frac{1}{1-e^s}$$

$$= \frac{1}{R}\left[\frac{1}{s} - \frac{L/R}{s} + \frac{L/R}{s+R/L}\right] + \frac{1}{R}\left(\frac{1}{s} - \frac{1}{s+R/L}\right)\left(1 + e^s + e^{2s} + e^{3s} + \cdots\right).$$

Thus
$$i(t) = \frac{1}{R}\left(t - \frac{L}{R} + \frac{L}{R}e^{-Rt/L}\right) + \frac{1}{R}\sum_{n=1}^{\infty}\left(1 - e^{-R(t-n)/L}\right)\mathcal{U}(t-n).$$

For $0 \le t < 2$ we have

$$i(t) = \frac{1}{R}\left(t - \frac{L}{R} + \frac{L}{R}e^{-Rt/L}\right) + \frac{1}{R}\left(1 - e^{-R(t-1)/L}\right)\mathcal{U}(t-1)$$

$$= \begin{cases} \dfrac{1}{R}\left(t - \dfrac{L}{R} + \dfrac{L}{R}e^{-Rt/L}\right), & 0 \le t < 1 \\ \dfrac{1}{R}\left(t - \dfrac{L}{R} + \dfrac{L}{R}e^{-Rt/L}\right) + \dfrac{1}{R}\left(1 - e^{-R(t-1)/L}\right), & 1 \le t < 2. \end{cases}$$

Exercises 5.4

12. The differential equation is $x'' + 2x' + x = 5f(t)$, where $f(t)$ is the square wave function with $a = \pi$. Using the initial conditions $x(0) = x'(0) = 0$ and taking the Laplace transform, we obtain

$$(s^2 + 2s + 1)\mathcal{L}\{x(t)\} = \frac{5}{s}\frac{1}{1+e^{-\pi s}} = \frac{5}{s}(1 - e^{-\pi s} + e^{-2\pi s} - e^{-3\pi s} + e^{-4\pi s} - \cdots)$$

$$= \frac{5}{s}\sum_{n=0}^{\infty}(-1)^n e^{-n\pi s}.$$

Then

$$\mathcal{L}\{x(t)\} = \frac{5}{s(s+1)^2}\sum_{n=0}^{\infty}(-1)^n e^{-n\pi s} = 5\sum_{n=0}^{\infty}(-1)^n\left(\frac{1}{s} - \frac{1}{s+1} - \frac{1}{(s+1)^2}\right)e^{-n\pi s}$$

and

$$x(t) = 5\sum_{n=0}^{\infty}(-1)^n(1 - e^{-(t-n\pi)} - (t-n\pi)e^{-(t-n\pi)})\,\mathcal{U}(t - n\pi).$$

The graph of $x(t)$ on the interval $[0, 4\pi)$ is shown below.

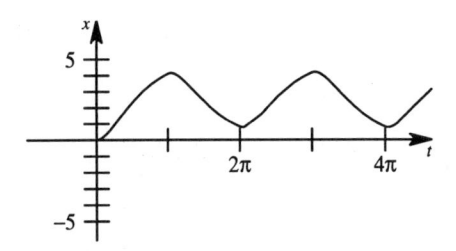

Exercises 5.5

3. The Laplace transform of the differential equation yields

$$\mathcal{L}\{y\} = \frac{1}{s^2 + 1}\left(1 + e^{-2\pi s}\right)$$

so that

$$y = \sin t + \sin t\,\mathcal{U}(t - 2\pi).$$

6. The Laplace transform of the differential equation yields

$$\mathcal{L}\{y\} = \frac{s}{s^2+1} + \frac{1}{s^2+1}(e^{-2\pi s} + e^{-4\pi s})$$

so that

$$y = \cos t + \sin t[\mathcal{U}(t - 2\pi) + \mathcal{U}(t - 4\pi)].$$

9. The Laplace transform of the differential equation yields

$$\mathcal{L}\{y\} = \frac{1}{(s+2)^2 + 1}e^{-2\pi s}$$

so that
$$y = e^{-2(t-2\pi)} \sin t\, \mathcal{U}(t - 2\pi).$$

12. The Laplace transform of the differential equation yields

$$\mathcal{L}\{y\} = \frac{1}{(s-1)^2(s-6)} + \frac{e^{-2s} + e^{-4s}}{(s-1)(s-6)}$$

$$= -\frac{1}{25}\frac{1}{s-1} - \frac{1}{5}\frac{1}{(s-1)^2} + \frac{1}{25}\frac{1}{s-6} + \left[-\frac{1}{5}\frac{1}{s-1} + \frac{1}{5}\frac{1}{s-6}\right](e^{-2s} + e^{-4s})$$

so that
$$y = -\frac{1}{25}e^t - \frac{1}{5}te^t + \frac{1}{25}e^{6t} + \left[-\frac{1}{5}e^{t-2} + \frac{1}{5}e^{6(t-2)}\right]\mathcal{U}(t-2)$$
$$+ \left[-\frac{1}{5}e^{t-4} + \frac{1}{5}e^{6(t-4)}\right]\mathcal{U}(t-4).$$

15. Using the sifting property, we have

$$\mathcal{L}\{\cos t\, \delta(t - 3\pi)\} = \int_0^\infty e^{-st} \cos t\, \delta(t - 3\pi)\, dt = e^{-st} \cos t\, \Big|_{t=3\pi} = -e^{-3\pi s}.$$

The Laplace transform of the differential equation is then

$$s^2 \mathcal{L}\{y\} - s + 1 + 2s\mathcal{L}\{y\} - 2 + 2\mathcal{L}\{y\} = -e^{-3\pi s}$$

so that

$$\mathcal{L}\{y\} = \frac{s+1}{s^2+2s+2} - \frac{1}{s^2+2s+2}e^{-3\pi s} = \frac{s+1}{(s+1)^2+1} - \frac{1}{(s+1)^2+1}e^{-3\pi s}$$

and

$$y = e^{-t}\cos t - e^{-(t-3\pi)}\sin(t-3\pi)\,\mathcal{U}(t-3\pi) = e^{-t}\cos t + e^{-(t-3\pi)}\sin t\,\mathcal{U}(t-3\pi).$$

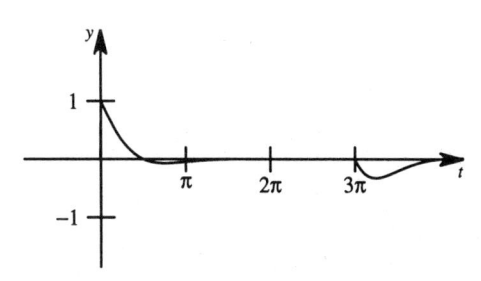

Exercises 5.6

3. Taking the Laplace transform of the system gives
$$s\mathcal{L}\{x\} + 1 = \mathcal{L}\{x\} - 2\mathcal{L}\{y\}$$
$$s\mathcal{L}\{y\} - 2 = 5\mathcal{L}\{x\} - \mathcal{L}\{y\}$$

so that
$$\mathcal{L}\{x\} = \frac{-s-5}{s^2+9} = -\frac{s}{s^2+9} - \frac{5}{3}\frac{3}{s^2+9}$$

and
$$x = -\cos 3t - \frac{5}{3}\sin 3t.$$

Then
$$y = \frac{1}{2}x - \frac{1}{2}x' = 2\cos 3t - \frac{7}{3}\sin 3t.$$

6. Taking the Laplace transform of the system gives
$$(s+1)\mathcal{L}\{x\} - (s-1)\mathcal{L}\{y\} = -1$$
$$s\mathcal{L}\{x\} + (s+2)\mathcal{L}\{y\} = 1$$

so that
$$\mathcal{L}\{y\} = \frac{s+1/2}{s^2+s+1} = \frac{s+1/2}{(s+1/2)^2 + (\sqrt{3}/2)^2}$$

and
$$\mathcal{L}\{x\} = \frac{-3/2}{s^2+s+1} = \frac{-3/2}{(s+1/2)^2 + (\sqrt{3}/2)^2}.$$

Then
$$y = e^{-t/2}\cos\frac{\sqrt{3}}{2}t \quad \text{and} \quad x = e^{-t/2}\sin\frac{\sqrt{3}}{2}t.$$

9. Adding the equations and then subtracting them gives
$$\frac{d^2x}{dt^2} = \frac{1}{2}t^2 + 2t$$
$$\frac{d^2y}{dt^2} = \frac{1}{2}t^2 - 2t.$$

Taking the Laplace transform of the system gives
$$\mathcal{L}\{x\} = 8\frac{1}{s} + \frac{1}{24}\frac{4!}{s^5} + \frac{1}{3}\frac{3!}{s^4}$$

and
$$\mathcal{L}\{y\} = \frac{1}{24}\frac{4!}{s^5} - \frac{1}{3}\frac{3!}{s^4}$$

so that
$$x = 8 + \frac{1}{24}t^4 + \frac{1}{3}t^3 \quad \text{and} \quad y = \frac{1}{24}t^4 - \frac{1}{3}t^3.$$

Exercises 5.6

12. Taking the Laplace transform of the system gives

$$(s-4)\mathcal{L}\{x\} + 2\mathcal{L}\{y\} = \frac{2e^{-s}}{s}$$

$$-3\mathcal{L}\{x\} + (s+1)\mathcal{L}\{y\} = \frac{1}{2} + \frac{e^{-s}}{s}$$

so that

$$\mathcal{L}\{x\} = \frac{-1/2}{(s-1)(s-2)} + e^{-s}\frac{1}{(s-1)(s-2)}$$

$$= \left[\frac{1}{2}\frac{1}{s-1} - \frac{1}{2}\frac{1}{s-2}\right] + e^{-s}\left[-\frac{1}{s-1} + \frac{1}{s-2}\right]$$

and

$$\mathcal{L}\{y\} = \frac{e^{-s}}{s} + \frac{s/4 - 1}{(s-1)(s-2)} + e^{-s}\frac{-s/2 + 2}{(s-1)(s-2)}$$

$$= \frac{3}{4}\frac{1}{s-1} - \frac{1}{2}\frac{1}{s-2} + e^{-s}\left[\frac{1}{s} - \frac{3}{2}\frac{1}{s-1} + \frac{1}{s-2}\right].$$

Then

$$x = \frac{1}{2}e^t - \frac{1}{2}e^{2t} + \left[-e^{t-1} + e^{2(t-1)}\right]\mathcal{U}(t-1)$$

and

$$y = \frac{3}{4}e^t - \frac{1}{2}e^{2t} + \left[1 - \frac{3}{2}e^{t-1} + e^{2(t-1)}\right]\mathcal{U}(t-1).$$

15. (a) By Kirchoff's first law we have $i_1 = i_2 + i_3$. By Kirchoff's second law, on each loop we have $E(t) = Ri_1 + L_1 i_2'$ and $E(t) = Ri_1 + L_2 i_3'$ or $L_1 i_2' + Ri_2 + Ri_3 = E(t)$ and $L_2 i_3' + Ri_2 + Ri_3 = E(t)$.

(b) Taking the Laplace transform of the system

$$0.01 i_2' + 5i_2 + 5i_3 = 100$$

$$0.0125 i_3' + 5i_2 + 5i_3 = 100$$

gives

$$(s+500)\mathcal{L}\{i_2\} + 500\mathcal{L}\{i_3\} = \frac{10{,}000}{s}$$

$$400\mathcal{L}\{i_2\} + (s+400)\mathcal{L}\{i_3\} = \frac{8{,}000}{s}$$

so that

$$\mathcal{L}\{i_3\} = \frac{8{,}000}{s^2 + 900s} = \frac{80}{9}\frac{1}{s} - \frac{80}{9}\frac{1}{s+900}.$$

Then

$$i_3 = \frac{80}{9} - \frac{80}{9}e^{-900t} \quad \text{and} \quad i_2 = 20 - 0.0025 i_3' - i_3 = \frac{100}{9} - \frac{100}{9}e^{-900t}.$$

(c) $i_1 = i_2 + i_3 = 20 - 20e^{-900t}$

71

Exercises 5.6

18. By Kirchoff's first law we have $i_1 = i_2 + i_3$. By Kirchoff's second law, on each loop we have $E(t) = Li_1' + Ri_2$ and $E(t) = Li_1' + \frac{1}{C}q$ so that $q = CRi_2$. Then $i_3 = q' = CRi_2'$ so that system is

$$Li' + Ri_2 = E(t)$$

$$CRi_2' + i_2 - i_1 = 0.$$

21. (a) Using Kirchoff's first law we write $i_1 = i_2 + i_3$. Since $i_2 = dq/dt$ we have $i_1 - i_3 = dq/dt$. Using Kirchoff's second law and summing the voltage drops across the shorter loop gives

$$E(t) = iR_1 + \frac{1}{C}q, \qquad (1)$$

so that

$$i_1 = \frac{1}{R_1}E(t) - \frac{1}{R_1 C}q.$$

Then

$$\frac{dq}{dt} = i_1 - i_3 = \frac{1}{R_1}E(t) - \frac{1}{R_1 C}q - i_3$$

and

$$R_1\frac{dq}{dt} + \frac{1}{C}q + R_1 i_3 = E(t).$$

Summing the voltage drops across the longer loop gives

$$E(t) = i_1 R_1 + L\frac{di_3}{dt} + R_2 i_3.$$

Combining this with (1) we obtain

$$i_1 R_1 + L\frac{di_3}{dt} + R_2 i_3 = i_1 R_1 + \frac{1}{C}q$$

or

$$L\frac{di_3}{dt} + R_2 i_3 - \frac{1}{C}q = 0.$$

(b) Using $L = R_1 = R_2 = C = 1$, $E(t) = 50e^{-t}\mathcal{U}(t-1) = 50e^{-1}e^{-(t-1)}\mathcal{U}(t-1)$, $q(0) = i_3(0) = 0$, and taking the Laplace transform of the system we obtain

$$(s+1)\mathcal{L}\{q\} + \mathcal{L}\{i_3\} = \frac{50e^{-1}}{s+1}e^{-s}$$

$$(s+1)\mathcal{L}\{i_3\} - \mathcal{L}\{q\} = 0,$$

so that

$$\mathcal{L}\{q\} = \frac{50e^{-1}e^{-s}}{(s+1)^2 + 1}$$

and

$$q(t) = 50e^{-1}e^{-(t-1)}\sin(t-1)\mathcal{U}(t-1) = 50e^{-t}\sin(t-1)\mathcal{U}(t-1).$$

6 Series Solutions

Exercises 6.1

3. Substituting $y = \sum_{n=0}^{\infty} c_n x^n$ into the differential equation we have

$$y'' - 2xy' + y = \underbrace{\sum_{n=2}^{\infty} n(n-1)c_n x^{n-2}}_{k=n-2} - 2\underbrace{\sum_{n=1}^{\infty} nc_n x^n}_{k=n} + \underbrace{\sum_{n=0}^{\infty} c_n x^n}_{k=n}$$

$$= \sum_{k=0}^{\infty}(k+2)(k+1)c_{k+2}x^k - 2\sum_{k=1}^{\infty} kc_k x^k + \sum_{k=0}^{\infty} c_k x^k$$

$$= 2c_2 + c_0 + \sum_{k=1}^{\infty}[(k+2)(k+1)c_{k+2} - (2k-1)c_k]x^k = 0.$$

Thus
$$2c_2 + c_0 = 0$$
$$(k+2)(k+1)c_{k+2} - (2k-1)c_k = 0$$

and
$$c_2 = -\frac{1}{2}c_0$$

$$c_{k+2} = \frac{2k-1}{(k+2)(k+1)}c_k, \quad k = 1, 2, 3, \ldots.$$

Choosing $c_0 = 1$ and $c_1 = 0$ we find

$$c_2 = -\frac{1}{2}$$

$$c_3 = c_5 = c_7 = \cdots = 0$$

$$c_4 = -\frac{1}{8}$$

$$c_6 = -\frac{7}{240}$$

and so on. For $c_0 = 0$ and $c_1 = 1$ we obtain

$$c_2 = c_4 = c_6 = \cdots = 0$$

$$c_3 = \frac{1}{6}$$

$$c_5 = \frac{1}{24}$$

$$c_7 = \frac{1}{112}$$

73

Exercises 6.1

and so on. Thus, two solutions are

$$y_1 = 1 - \frac{1}{2}x^2 - \frac{1}{8}x^4 - \frac{7}{240}x^6 - \cdots \quad \text{and} \quad y_2 = x + \frac{1}{6}x^3 + \frac{1}{24}x^5 + \frac{1}{112}x^7 + \cdots.$$

6. Substituting $y = \sum_{n=0}^{\infty} c_n x^n$ into the differential equation we have

$$y'' + 2xy' + 2y = \underbrace{\sum_{n=2}^{\infty} n(n-1)c_n x^{n-2}}_{k=n-2} + 2\underbrace{\sum_{n=1}^{\infty} nc_n x^n}_{k=n} + 2\underbrace{\sum_{n=0}^{\infty} c_n x^n}_{k=n}$$

$$= \sum_{k=0}^{\infty}(k+2)(k+1)c_{k+2}x^k + 2\sum_{k=1}^{\infty} kc_k x^k + 2\sum_{k=0}^{\infty} c_k x^k$$

$$= 2c_2 + 2c_0 + \sum_{k=1}^{\infty}[(k+2)(k+1)c_{k+2} + 2(k+1)c_k]x^k = 0.$$

Thus
$$2c_2 + 2c_0 = 0$$
$$(k+2)(k+1)c_{k+2} + 2(k+1)c_k = 0$$

and
$$c_2 = -c_0$$
$$c_{k+2} = -\frac{2}{k+2}c_k, \quad k = 1, 2, 3, \ldots.$$

Choosing $c_0 = 1$ and $c_1 = 0$ we find

$$c_2 = -1$$
$$c_3 = c_5 = c_7 = \cdots = 0$$
$$c_4 = \frac{1}{2}$$
$$c_6 = -\frac{1}{6}$$

and so on. For $c_0 = 0$ and $c_1 = 1$ we obtain

$$c_2 = c_4 = c_6 = \cdots = 0$$
$$c_3 = -\frac{2}{3}$$
$$c_5 = \frac{4}{15}$$
$$c_7 = -\frac{8}{105}$$

and so on. Thus, two solutions are
$$y_1 = 1 - x^2 + \frac{1}{2}x^4 - \frac{1}{6}x^6 + \cdots \quad \text{and} \quad y_2 = x - \frac{2}{3}x^3 + \frac{4}{15}x^5 - \frac{8}{105}x^7 + \cdots.$$

9. Substituting $y = \sum_{n=0}^{\infty} c_n x^n$ into the differential equation we have

$$(x^2 - 1)y'' + 4xy' + 2y = \underbrace{\sum_{n=2}^{\infty} n(n-1)c_n x^n}_{k=n} - \underbrace{\sum_{n=2}^{\infty} n(n-1)c_n x^{n-2}}_{k=n-2} + 4\underbrace{\sum_{n=1}^{\infty} nc_n x^n}_{k=n} + 2\underbrace{\sum_{n=0}^{\infty} c_n x^n}_{k=n}$$

$$= \sum_{k=2}^{\infty} k(k-1)c_k x^k - \sum_{k=0}^{\infty} (k+2)(k+1)c_{k+2} x^k + 4\sum_{k=1}^{\infty} kc_k x^k + 2\sum_{k=0}^{\infty} c_k x^k$$

$$= -2c_2 + 2c_0 + (-6c_3 + 6c_1)x + \sum_{k=2}^{\infty} \left[\left(k^2 - k + 4k + 2\right)c_k - (k+2)(k+1)c_{k+2}\right]x^k = 0.$$

Thus
$$-2c_2 + 2c_0 = 0$$
$$-6c_3 + 6c_1 = 0$$
$$\left(k^2 + 3k + 2\right)c_k - (k+2)(k+1)c_{k+2} = 0$$

and
$$c_2 = c_0$$
$$c_3 = c_1$$
$$c_{k+2} = c_k, \quad k = 2, 3, 4, \ldots.$$

Choosing $c_0 = 1$ and $c_1 = 0$ we find
$$c_2 = 1$$
$$c_3 = c_5 = c_7 = \cdots = 0$$
$$c_4 = c_6 = c_8 = \cdots = 1.$$

For $c_0 = 0$ and $c_1 = 1$ we obtain
$$c_2 = c_4 = c_6 = \cdots = 0$$
$$c_3 = c_5 = c_7 = \cdots = 1.$$

Thus, two solutions are
$$y_1 = 1 + x^2 + x^4 + \cdots \quad \text{and} \quad y_2 = x + x^3 + x^5 + \cdots.$$

Exercises 6.1

12. Substituting $y = \sum_{n=0}^{\infty} c_n x^n$ into the differential equation we have

$$\left(x^2 - 1\right) y'' + xy' - y = \underbrace{\sum_{n=2}^{\infty} n(n-1)c_n x^n}_{k=n} - \underbrace{\sum_{n=2}^{\infty} n(n-1)c_n x^{n-2}}_{k=n-2} + \underbrace{\sum_{n=1}^{\infty} n c_n x^n}_{k=n} - \underbrace{\sum_{n=0}^{\infty} c_n x^n}_{k=n}$$

$$= \sum_{k=2}^{\infty} k(k-1)c_k x^k - \sum_{k=0}^{\infty} (k+2)(k+1)c_{k+2} x^k + \sum_{k=1}^{\infty} k c_k x^k - \sum_{k=0}^{\infty} c_k x^k$$

$$= (-c_2 - c_0) - 6c_3 x + \sum_{k=2}^{\infty} \left[-(k+2)(k+1)c_{k+2} + \left(k^2 - 1\right) c_k \right] x^k = 0.$$

Thus
$$-2c_2 - c_0 = 0$$

$$-6c_3 = 0$$

$$-(k+2)(k+1)c_{k+2} + (k-1)(k+1)c_k = 0$$

and
$$c_2 = -\frac{1}{2} c_0$$

$$c_3 = 0$$

$$c_{k+2} = \frac{k-1}{k+2} c_k, \quad k = 2, 3, 4, \ldots.$$

Choosing $c_0 = 1$ and $c_1 = 0$ we find

$$c_2 = -\frac{1}{2}$$

$$c_3 = c_5 = c_7 = \cdots = 0$$

$$c_4 = -\frac{1}{8}$$

and so on. For $c_0 = 0$ and $c_1 = 1$ we obtain

$$c_2 = c_4 = c_6 = \cdots = 0$$

$$c_3 = c_5 = c_7 = \cdots = 0.$$

Thus, two solutions are

$$y_1 = 1 - \frac{1}{2} x^2 - \frac{1}{8} x^4 - \cdots \quad \text{and} \quad y_2 = x.$$

15. Substituting $y = \sum_{n=0}^{\infty} c_n x^n$ into the differential equation we have

$$(x-1)y'' - xy' + y = \sum_{n=2}^{\infty} n(n-1)c_n x^{n-1} - \sum_{n=2}^{\infty} n(n-1)c_n x^{n-2} - \sum_{n=1}^{\infty} nc_n x^n + \sum_{n=0}^{\infty} c_n x^n$$

$$\underbrace{}_{k=n-1} \quad \underbrace{}_{k=n-2} \quad \underbrace{}_{k=n} \quad \underbrace{}_{k=n}$$

$$= \sum_{k=1}^{\infty} (k+1)kc_{k+1}x^k - \sum_{k=0}^{\infty} (k+2)(k+1)c_{k+2}x^k - \sum_{k=1}^{\infty} kc_k x^k + \sum_{k=0}^{\infty} c_k x^k$$

$$= -2c_2 + c_0 + \sum_{k=1}^{\infty} [-(k+2)(k+1)c_{k+2} + (k+1)kc_{k+1} - (k-1)c_k]x^k = 0.$$

Thus
$$-2c_2 + c_0 = 0$$

$$-(k+2)(k+1)c_{k+2} + (k-1)kc_{k+1} - (k-1)c_k = 0$$

and
$$c_2 = \frac{1}{2}c_0$$

$$c_{k+2} = \frac{kc_{k+1}}{k+2} - \frac{(k-1)c_k}{(k+2)(k+1)}, \quad k = 1, 2, 3, \ldots.$$

Choosing $c_0 = 1$ and $c_1 = 0$ we find

$$c_2 = \frac{1}{2}, \quad c_3 = \frac{1}{6}, \quad c_4 = 0$$

and so on. For $c_0 = 0$ and $c_1 = 1$ we obtain $c_2 = c_3 = c_4 = \cdots = 0$. Thus,

$$y = C_1\left(1 + \frac{1}{2}x^2 + \frac{1}{6}x^3 + \cdots\right) + C_2 x$$

and
$$y' = C_1\left(x + \frac{1}{2}x^2 + \cdots\right) + C_2.$$

The initial conditions imply $C_1 = -2$ and $C_2 = 6$, so

$$y = -2\left(1 + \frac{1}{2}x^2 + \frac{1}{6}x^3 + \cdots\right) + 6x = 8x - 2e^x.$$

Exercises 6.1

18. Substituting $y = \sum_{n=0}^{\infty} c_n x^n$ into the differential equation we have

$$(x^2+1)y'' + 2xy' = \underbrace{\sum_{n=2}^{\infty} n(n-1)c_n x^n}_{k=n} + \underbrace{\sum_{n=2}^{\infty} n(n-1)c_n x^{n-2}}_{k=n-2} + \underbrace{\sum_{n=1}^{\infty} 2nc_n x^n}_{k=n}$$

$$= \sum_{k=2}^{\infty} k(k-1)c_k x^k + \sum_{k=0}^{\infty} (k+2)(k+1)c_{k+2} x^k + \sum_{k=1}^{\infty} 2kc_k x^k$$

$$= 2c_2 + (6c_3 + 2c_1)x + \sum_{k=2}^{\infty} [k(k+1)c_k + (k+2)(k+1)c_{k+2}]x^k = 0.$$

Thus
$$2c_2 = 0$$
$$6c_3 + 2c_1 = 0$$
$$k(k+1)c_k + (k+2)(k+1)c_{k+2} = 0$$

and
$$c_2 = 0$$
$$c_3 = -\frac{1}{3}c_1$$
$$c_{k+2} = -\frac{k}{k+2}c_k, \quad k = 2, 3, 4, \ldots.$$

Choosing $c_0 = 1$ and $c_1 = 0$ we find $c_3 = c_4 = c_5 = \cdots = 0$. For $c_0 = 0$ and $c_1 = 1$ we obtain

$$c_3 = -\frac{1}{3}$$
$$c_4 = c_6 = c_8 = \cdots = 0$$
$$c_5 = -\frac{1}{5}$$
$$c_7 = \frac{1}{7}$$

and so on. Thus
$$y = c_0 + c_1\left(x - \frac{1}{3}x^3 + \frac{1}{5}x^5 - \frac{1}{7}x^7 + \cdots\right)$$

and
$$y' = c_1\left(1 - x^2 + x^4 - x^6 + \cdots\right).$$

Exercises 6.1

The initial conditions imply $c_0 = 0$ and $c_1 = 1$, so
$$y = x - \frac{1}{3}x^3 + \frac{1}{5}x^5 - \frac{1}{7}x^7 + \cdots.$$

21. Substituting $y = \sum_{n=0}^{\infty} c_n x^n$ into the differential equation we have

$$y'' + e^{-x}y = \sum_{n=2}^{\infty} n(n-1)c_n x^{n-2}$$

$$+ \left(1 - x + \frac{1}{2}x^2 - \frac{1}{6}x^3 + \frac{1}{24}x^4 - \cdots\right)\left(c_0 + c_1 x + c_2 x^2 + c_3 x^3 + \cdots\right)$$

$$= \left[2c_2 + 6c_3 x + 12c_4 x^2 + 20c_5 x^3 + \cdots\right] + \left[c_0 + (c_1 - c_0)x + \left(c_2 - c_1 + \frac{1}{2}c_0\right)x^2 + \cdots\right]$$

$$= (2c_2 + c_0) + (6c_3 + c_1 - c_0)x + \left(12c_4 + c_2 - c_1 + \frac{1}{2}c_0\right)x^2 + \cdots = 0.$$

Thus
$$2c_2 + c_0 = 0$$
$$6c_3 + c_1 - c_0 = 0$$
$$12c_4 + c_2 - c_1 + \frac{1}{2}c_0 = 0$$

and
$$c_2 = -\frac{1}{2}c_0$$
$$c_3 = -\frac{1}{6}c_1 + \frac{1}{6}c_0$$
$$c_4 = -\frac{1}{12}c_2 + \frac{1}{12}c_1 - \frac{1}{24}c_0.$$

Choosing $c_0 = 1$ and $c_1 = 0$ we find
$$c_2 = -\frac{1}{2}, \quad c_3 = \frac{1}{6}, \quad c_4 = 0$$
and so on. For $c_0 = 0$ and $c_1 = 1$ we obtain
$$c_2 = 0, \quad c_3 = -\frac{1}{6}, \quad c_4 = \frac{1}{12}.$$

Thus, two solutions are
$$y_1 = 1 - \frac{1}{2}x^2 + \frac{1}{6}x^3 + \cdots \quad \text{and} \quad y_2 = x - \frac{1}{6}x^3 + \frac{1}{12}x^4 + \cdots.$$

Exercises 6.2

3. Irregular singular point: $x = 3$; regular singular point: $x = -3$

6. Irregular singular point: $x = 5$; regular singular point: $x = 0$

9. Irregular singular point: $x = 0$; regular singular points: $x = 2, \pm 5$

12. Substituting $y = \sum_{n=0}^{\infty} c_n x^{n+r}$ into the differential equation and collecting terms, we obtain

$$2xy'' + 5y' + xy = \left(2r^2 + 3r\right)c_0 x^{r-1} + \left(2r^2 + 7r + 5\right)c_1 x^r$$

$$+ \sum_{k=2}^{\infty} [2(k+r)(k+r-1)c_k + 5(k+r)c_k + c_{k-2}]x^{k+r-1}$$

$$= 0,$$

which implies
$$2r^2 + 3r = r(2r+3) = 0,$$
$$\left(2r^2 + 7r + 5\right)c_1 = 0,$$

and
$$(k+r)(2k+2r+3)c_k + c_{k-2} = 0.$$

The indicial roots are $r = -3/2$ and $r = 0$, so $c_1 = 0$. For $r = -3/2$ the recurrence relation is

$$c_k = -\frac{c_{k-2}}{(2k-3)k}, \quad k = 2, 3, 4, \ldots,$$

and
$$c_2 = -\frac{1}{2}c_0, \quad c_3 = 0, \quad c_4 = \frac{1}{40}c_0.$$

For $r = 0$ the recurrence relation is

$$c_k = -\frac{c_{k-2}}{k(2k+3)}, \quad k = 2, 3, 4, \ldots,$$

and
$$c_2 = -\frac{1}{14}c_0, \quad c_3 = 0, \quad c_4 = \frac{1}{616}c_0.$$

The general solution on $(0, \infty)$ is

$$y = C_1 x^{-3/2}\left(1 - \frac{1}{2}x^2 + \frac{1}{40}x^4 + \cdots\right) + C_2\left(1 - \frac{1}{14}x^2 + \frac{1}{616}x^4 + \cdots\right).$$

15. Substituting $y = \sum_{n=0}^{\infty} c_n x^{n+r}$ into the differential equation and collecting terms, we obtain

$$3xy'' + (2-x)y' - y = \left(3r^2 - r\right)c_0 x^{r-1}$$

$$+ \sum_{k=1}^{\infty} [3(k+r-1)(k+r)c_k + 2(k+r)c_k - (k+r)c_{k-1}]x^{k+r-1}$$

$$= 0,$$

80

Exercises 6.2

which implies
$$3r^2 - r = r(3r-1) = 0$$

and
$$(k+r)(3k+3r-1)c_k - (k+r)c_{k-1} = 0.$$

The indicial roots are $r = 0$ and $r = 1/3$. For $r = 0$ the recurrence relation is

$$c_k = \frac{c_{k-1}}{(3k-1)}, \quad k = 1, 2, 3, \ldots,$$

and
$$c_1 = \frac{1}{2}c_0, \quad c_2 = \frac{1}{10}c_0, \quad c_3 = \frac{1}{80}c_0.$$

For $r = 1/3$ the recurrence relation is

$$c_k = \frac{c_{k-1}}{3k}, \quad k = 1, 2, 3, \ldots,$$

and
$$c_1 = \frac{1}{3}c_0, \quad c_2 = \frac{1}{18}c_0, \quad c_3 = \frac{1}{162}c_0.$$

The general solution on $(0, \infty)$ is

$$y = C_1 \left(1 + \frac{1}{2}x + \frac{1}{10}x^2 + \frac{1}{80}x^3 + \cdots \right) + C_2 x^{1/3} \left(1 + \frac{1}{3}x + \frac{1}{18}x^2 + \frac{1}{162}x^3 + \cdots \right).$$

18. Substituting $y = \sum_{n=0}^{\infty} c_n x^{n+r}$ into the differential equation and collecting terms, we obtain

$$x^2 y'' + xy' + \left(x^2 - \frac{4}{9}\right) y = \left(r^2 - \frac{4}{9}\right) c_0 x^r + \left(r^2 + 2r + \frac{5}{9}\right) c_1 x^{r+1}$$

$$+ \sum_{k=2}^{\infty} \left[(k+r)(k+r-1)c_k + (k+r)c_k - \frac{4}{9}c_k + c_{k-2}\right] x^{k+r}$$

$$= 0,$$

which implies
$$r^2 - \frac{4}{9} = \left(r + \frac{2}{3}\right)\left(r - \frac{2}{3}\right) = 0,$$

$$\left(r^2 + 2r + \frac{5}{9}\right) c_1 = 0,$$

and
$$\left[(k+r)^2 - \frac{4}{9}\right] c_k + c_{k-2} = 0.$$

The indicial roots are $r = -2/3$ and $r = 2/3$, so $c_1 = 0$. For $r = -2/3$ the recurrence relation is

$$c_k = -\frac{9c_{k-2}}{3k(3k-4)}, \quad k = 2, 3, 4, \ldots,$$

and
$$c_2 = -\frac{3}{4}c_0, \quad c_3 = 0, \quad c_4 = \frac{9}{128}c_0.$$

81

Exercises 6.2

For $r = 2/3$ the recurrence relation is

$$c_k = -\frac{9c_{k-2}}{3k(3k+4)}, \quad k = 2, 3, 4, \ldots,$$

and

$$c_2 = -\frac{3}{20}c_0, \quad c_3 = 0, \quad c_4 = \frac{9}{1{,}280}c_0.$$

The general solution on $(0, \infty)$ is

$$y = C_1 x^{-2/3}\left(1 - \frac{3}{4}x^2 + \frac{9}{128}x^4 + \cdots\right) + C_2 x^{2/3}\left(1 - \frac{3}{20}x^2 + \frac{9}{1{,}280}x^4 + \cdots\right).$$

21. Substituting $y = \sum_{n=0}^{\infty} c_n x^{n+r}$ into the differential equation and collecting terms, we obtain

$$2x^2 y'' - x(x-1)y' - y = \left(2r^2 - r - 1\right)c_0 x^r$$

$$+ \sum_{k=1}^{\infty}[2(k+r)(k+r-1)c_k + (k+r)c_k - c_k - (k+r-1)c_{k-1}]x^{k+r}$$

$$= 0,$$

which implies

$$2r^2 - r - 1 = (2r+1)(r-1) = 0$$

and

$$[(k+r)(2k+2r-1) - 1]c_k - (k+r-1)2c_{k-1} = 0.$$

The indicial roots are $r = -1/2$ and $r = 1$. For $r = -1/2$ the recurrence relation is

$$c_k = \frac{c_{k-1}}{2k}, \quad k = 1, 2, 3, \ldots,$$

and

$$c_1 = \frac{1}{2}c_0, \quad c_2 = \frac{1}{8}c_0, \quad c_3 = \frac{1}{48}c_0.$$

For $r = 1$ the recurrence relation is

$$c_k = \frac{c_{k-1}}{2k+3}, \quad k = 1, 2, 3, \ldots,$$

and

$$c_1 = \frac{1}{5}c_0, \quad c_2 = \frac{1}{35}c_0, \quad c_3 = \frac{1}{315}c_0.$$

The general solution on $(0, \infty)$ is

$$y = C_1 x^{-1/2}\left(1 + \frac{1}{2}x + \frac{1}{8}x^2 + \frac{1}{48}x^3 + \cdots\right) + C_2 x\left(1 + \frac{1}{5}x + \frac{1}{35}x^2 + \frac{1}{315}x^3 + \cdots\right).$$

24. Substituting $y = \sum_{n=0}^{\infty} c_n x^{n+r}$ into the differential equation and collecting terms, we obtain

$$x^2 y'' + xy' + \left(x^2 - \frac{1}{4}\right)y = \left(r^2 - \frac{1}{4}\right)c_0 x^r + \left(r^2 + 2r + \frac{3}{4}\right)c_1 x^{r+1}$$

$$+ \sum_{k=2}^{\infty}\left[(k+r)(k+r-1)c_k + (k+r)c_k - \frac{1}{4}c_k + c_{k-2}\right]x^{k+r}$$

$$= 0,$$

which implies
$$r^2 - \frac{1}{4} = \left(r - \frac{1}{2}\right)\left(r + \frac{1}{2}\right) = 0,$$

$$\left(r^2 + 2r + \frac{3}{4}\right)c_1 = 0,$$

and
$$\left[(k+r)^2 - \frac{1}{4}\right]c_k + c_{k-2} = 0.$$

The indicial roots are $r_1 = 1/2$ and $r_2 = -1/2$, so $c_1 = 0$. For $r_1 = 1/2$ the recurrence relation is
$$c_k = -\frac{c_{k-2}}{k(k+1)}, \quad k = 2, 3, 4, \ldots,$$

and
$$c_2 = -\frac{1}{3!}c_0$$
$$c_3 = c_5 = c_7 = \cdots = 0$$
$$c_4 = \frac{1}{5!}c_0$$
$$c_{2n} = \frac{(-1)^n}{(2n+1)!}c_0.$$

For $r_2 = -1/2$ the recurrence relation is
$$c_k = -\frac{c_{k-2}}{k(k-1)}, \quad k = 2, 3, 4, \ldots,$$

and
$$c_2 = -\frac{1}{2!}c_0$$
$$c_3 = c_5 = c_7 = \cdots = 0$$
$$c_4 = \frac{1}{4!}c_0$$
$$c_{2n} = \frac{(-1)^n}{(2n)!}c_0.$$

The general solution on $(0, \infty)$ is
$$y = C_1 x^{1/2} \sum_{n=0}^{\infty} \frac{(-1)^n}{(2n+1)!}x^{2n} + C_2 x^{-1/2} \sum_{n=0}^{\infty} \frac{(-1)^n}{(2n)!}x^{2n}$$

$$= C_1 x^{-1/2} \sum_{n=0}^{\infty} \frac{(-1)^n}{(2n+1)!}x^{2n+1} + C_2 x^{-1/2} \sum_{n=0}^{\infty} \frac{(-1)^n}{(2n)!}x^{2n}$$

$$= x^{-1/2}[C_1 \sin x + C_2 \cos x].$$

27. Substituting $y = \sum_{n=0}^{\infty} c_n x^{n+r}$ into the differential equation and collecting terms, we obtain
$$xy'' + (1-x)y' - y = r^2 c_0 x^{r-1} + \sum_{k=0}^{\infty}[(k+r)(k+r-1)c_k + (k+r)c_k - (k+r)c_{k-1}]x^{k+r-1} = 0,$$

83

Exercises 6.2

which implies $r^2 = 0$ and
$$(k+r)^2 c_k - (k+r)c_{k-1} = 0.$$
The indicial roots are $r_1 = r_2 = 0$ and the recurrence relation is
$$c_k = \frac{c_{k-1}}{k}, \quad k = 1, 2, 3, \ldots.$$
One solution is
$$y_1 = c_0 \left(1 + x + \frac{1}{2}x^2 + \frac{1}{3!}x^3 + \cdots\right) = c_0 e^x.$$
A second solution is
$$y_2 = y_1 \int \frac{e^{-\int (1/x - 1) dx}}{e^{2x}} dx = e^x \int \frac{e^x/x}{e^{2x}} dx = e^x \int \frac{1}{x} e^{-x} dx$$
$$= e^x \int \frac{1}{x}\left(1 - x + \frac{1}{2}x^2 - \frac{1}{3!}x^3 + \cdots\right) dx = e^x \int \left(\frac{1}{x} - 1 + \frac{1}{2}x - \frac{1}{3!}x^2 + \cdots\right) dx$$
$$= e^x \left[\ln x - x + \frac{1}{2 \cdot 2}x^2 - \frac{1}{3 \cdot 3!}x^3 + \cdots\right] = e^x \ln x - e^x \sum_{n=1}^{\infty} \frac{(-1)^{n+1}}{n \cdot n!} x^n.$$
The general solution on $(0, \infty)$ is
$$y = C_1 e^x + C_2 e^x \left(\ln x - \sum_{n=1}^{\infty} \frac{(-1)^{n+1}}{n \cdot n!} x^n\right).$$

30. Substituting $y = \sum_{n=0}^{\infty} c_n x^{n+r}$ into the differential equation and collecting terms, we obtain
$$xy'' - xy' + y = \left(r^2 - r\right) c_0 x^{r-1} + \sum_{k=0}^{\infty}[(k+r+1)(k+r)c_{k+1} - (k+r)c_k + c_k]x^{k+r} = 0$$
which implies
$$r^2 - r = r(r-1) = 0$$
and
$$(k+r+1)(k+r)c_{k+1} - (k+r-1)c_k = 0.$$
The indicial roots are $r_1 = 1$ and $r_2 = 0$. For $r_1 = 1$ the recurrence relation is
$$c_{k+1} = \frac{kc_k}{(k+2)(k+1)}, \quad k = 0, 1, 2, \ldots,$$
and one solution is $y_1 = c_0 x$. A second solution is
$$y_2 = x \int \frac{e^{-\int -dx}}{x^2} dx = x \int \frac{e^x}{x^2} dx = x \int \frac{1}{x^2}\left(1 + x + \frac{1}{2}x^2 + \frac{1}{3!}x^3 + \cdots\right) dx$$
$$= x \int \left(\frac{1}{x^2} + \frac{1}{x} + \frac{1}{2} + \frac{1}{3!}x + \frac{1}{4!}x^2 + \cdots\right) dx = x\left[-\frac{1}{x} + \ln x + \frac{1}{2}x + \frac{1}{12}x^2 + \frac{1}{72}x^3 + \cdots\right]$$
$$= x \ln x - 1 + \frac{1}{2}x^2 + \frac{1}{12}x^3 + \frac{1}{72}x^4 + \cdots.$$

The general solution on $(0, \infty)$ is
$$y = C_1 x + C_2 y_2(x).$$

Exercises 6.3

3. Since $\nu^2 = 25/4$ the general solution is $y = c_1 J_{5/2}(x) + c_2 J_{-5/2}(x)$.

6. Since $\nu^2 = 4$ the general solution is $y = c_1 J_2(x) + c_2 Y_2(x)$.

9. If $y = x^{-1/2} w(x)$ then
$$y' = x^{-1/2} w'(x) - \frac{1}{2} x^{-3/2} w(x),$$
$$y'' = x^{-1/2} w''(x) - x^{-3/2} w'(x) + \frac{3}{4} x^{-5/2} w(x),$$

and
$$x^2 y'' + 2xy' + \lambda^2 x^2 y = x^{3/2} w'' + x^{1/2} w' + \left(\lambda^2 x^{3/2} - \frac{1}{4} x^{-1/2}\right) w.$$

Multiplying by $x^{1/2}$ we obtain
$$x^2 w'' + x w' + \left(\lambda^2 x^2 - \frac{1}{4}\right) w = 0,$$
whose solution is $w = c_1 J_{1/2}(\lambda x) + c_2 J_{-1/2}(\lambda x)$. Then
$$y = c_1 x^{-1/2} J_{1/2}(\lambda x) + c_2 x^{-1/2} J_{-1/2}(\lambda x).$$

12. From $y = \sqrt{x}\, J_\nu(\lambda x)$ we find
$$y' = \lambda \sqrt{x}\, J'_\nu(\lambda x) + \frac{1}{2} x^{-1/2} J_\nu(\lambda x)$$
and
$$y'' = \lambda^2 \sqrt{x}\, J''_\nu(\lambda x) + \lambda x^{-1/2} J'_\nu(\lambda x) - \frac{1}{4} x^{-3/2} J_\nu(\lambda x).$$

Substituting into the differential equation, we have
$$x^2 y'' + \left(\lambda^2 x^2 - \nu^2 + \frac{1}{4}\right) y = \sqrt{x}\left[\lambda^2 x^2 J''_\nu(\lambda x) + \lambda x J'_\nu(\lambda x) + \left(\lambda^2 x^2 - \nu^2\right) J_\nu(\lambda x)\right]$$
$$= \sqrt{x} \cdot 0 \quad \text{(since } J_n \text{ is a solution of Bessel's equation)}$$
$$= 0.$$

Therefore, $\sqrt{x}\, J_\nu(\lambda x)$ is a solution of the original equation.

15. From Problem 10 with $n = -1$ we find $y = x^{-1} J_{-1}(x)$. From Problem 11 with $n = 1$ we find $y = x^{-1} J_1(x) = -x^{-1} J_{-1}(x)$.

Exercises 6.3

18. From Problem 10 with $n = 3$ we find $y = x^3 J_3(x)$. From Problem 11 with $n = -3$ we find $y = x^3 J_{-3}(x) = -x^3 J_3(x)$.

21. Write the recurrence relation in Problem 19 in the form

$$xJ'_\nu(x) - \nu J_\nu(x) = -xJ_{\nu+1}(x).$$

Dividing by x gives

$$J'_\nu(x) - \frac{\nu}{x}J_\nu(x) = -J_{\nu+1}(x).$$

This is a linear first-order differential equation in $J_\nu(x)$. Multiplying both sides of the equation by the integrating factor $x^{-\nu}$ gives

$$\frac{d}{dx}[x^{-\nu}J_\nu(x)] = -x^{-\nu}J_{\nu+1}(x).$$

24. (a) We identify $m = 4$, $k = 1$, and $\alpha = 0.1$. Then

$$x(t) = c_1 J_0(10e^{-0.05t}) + c_2 Y_0(10e^{-0.05t})$$

and

$$x'(t) = -0.5c_1 J'_0(10e^{-0.05t}) - 0.5c_2 Y'_0(10e^{-0.05t}).$$

Now $x(0) = 1$ and $x'(0) = -1/2$ imply

$$c_1 J_0(10) + c_2 Y_0(10) = 1$$

$$c_1 J'_0(10) + c_2 Y'_0(10) = 1.$$

Using Cramer's rule we obtain

$$c_1 = \frac{Y'_0(10) - Y_0(10)}{J_0(10)Y'_0(10) - J'_0(10)Y_0(10)}$$

and

$$c_2 = \frac{J_0(10) - J'_0(10)}{J_0(10)Y'_0(10) - J'_0(10)Y_0(10)}.$$

Using $Y'_0 = -Y_1$ and $J'_0 = -J_1$ and Table 6.1 we find $c_1 = -4.7860$ and $c_2 = -3.1803$. Thus

$$x(t) = -4.7860 J_0(10e^{-0.05t}) - 3.1803 Y_0(10e^{-0.05t}).$$

(b) In Problem 43(a) of Section 3.6 the conjecture was that the restoring force on the spring would decay to the point at which the spring would be incapable of returning the mass to the equilibrium position. This is corroborated by the graph.

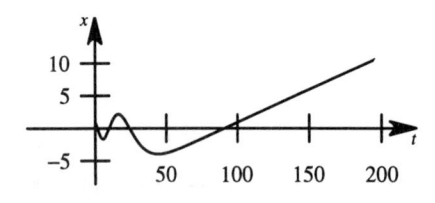

Exercises 6.3

27. (a) Identifying $\alpha = \frac{1}{2}$, the general solution of $x'' + \frac{1}{4}tx = 0$ is

$$x(t) = c_1 x^{1/2} J_{1/3}\left(\frac{1}{3}x^{3/2}\right) + c_2 x^{1/2} J_{-1/3}\left(\frac{1}{3}x^{3/2}\right).$$

Solving the system $x(0.1) = 1$, $x'(0.1) = -\frac{1}{2}$ we find $c_1 = -0.809264$ and $c_2 = 0.782397$.

(b) In Problem 43(b) of Section 3.6 the conjecture was that the oscillations would become periodic and the spring would oscillate more rapidly. This is corroborated by the graph.

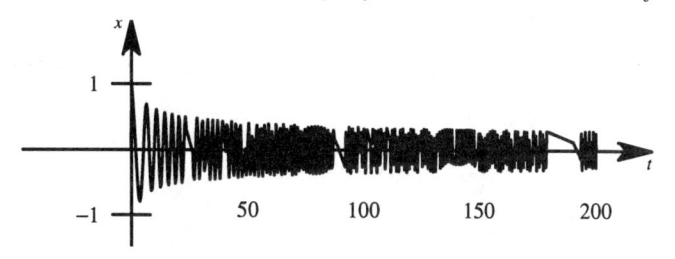

30. (a) Writing the differential equation in the form $xy'' + (PL/M)y = 0$, we identify $\lambda = PL/M$. From Problem 29 the solution of this differential equation is

$$y = c_1 \sqrt{x}\, J_1\left(2\sqrt{PLx/M}\right) + c_2 \sqrt{x}\, Y_1\left(2\sqrt{PLx/M}\right).$$

Now $J_1(0) = 0$, so $y(0) = 0$ implies $c_2 = 0$ and

$$y = c_1 \sqrt{x}\, J_1\left(2\sqrt{PLx/M}\right).$$

(b) From $y(L) = 0$ we have $y = J_1(2L\sqrt{PM}) = 0$. The first positive zero of J_1 is 3.8317 so, solving $2L\sqrt{P_1/M} = 3.8317$, we find $P_1 = 3.6705M/L^2$. Therefore,

$$y_1(x) = c_1 \sqrt{x}\, J_1\left(2\sqrt{\frac{3.6705x}{L}}\right) = c_1 \sqrt{x}\, J_1\left(\frac{3.8317}{L}\sqrt{x}\right).$$

(c) For $c_1 = 1$ and $L = 1$ the graph of $y_1 = \sqrt{x}\, J_1(3.8317\sqrt{x})$ is shown.

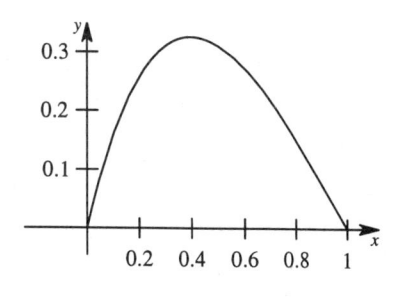

Appendix

Appendix I

3. (a) $AB = \begin{pmatrix} -2-9 & 12-6 \\ 5+12 & -30+8 \end{pmatrix} = \begin{pmatrix} -11 & 6 \\ 17 & -22 \end{pmatrix}$

   (b) $BA = \begin{pmatrix} -2-30 & 3+24 \\ 6-10 & -9+8 \end{pmatrix} = \begin{pmatrix} -32 & 27 \\ -4 & -1 \end{pmatrix}$

   (c) $A^2 = \begin{pmatrix} 4+15 & -6-12 \\ -10-20 & 15+16 \end{pmatrix} = \begin{pmatrix} 19 & -18 \\ -30 & 31 \end{pmatrix}$

   (d) $B^2 = \begin{pmatrix} 1+18 & -6+12 \\ -3+6 & 18+4 \end{pmatrix} = \begin{pmatrix} 19 & 6 \\ 3 & 22 \end{pmatrix}$

6. (a) $AB = \begin{pmatrix} 5 & -6 & 7 \end{pmatrix} \begin{pmatrix} 3 \\ 4 \\ -1 \end{pmatrix} = (-16)$

   (b) $BA = \begin{pmatrix} 3 \\ 4 \\ -1 \end{pmatrix} \begin{pmatrix} 5 & -6 & 7 \end{pmatrix} = \begin{pmatrix} 15 & -18 & 21 \\ 20 & -24 & 28 \\ -5 & 6 & -7 \end{pmatrix}$

   (c) $(BA)C = \begin{pmatrix} 15 & -18 & 21 \\ 20 & -24 & 28 \\ -5 & 6 & -7 \end{pmatrix} \begin{pmatrix} 1 & 2 & 4 \\ 0 & 1 & -1 \\ 3 & 2 & 1 \end{pmatrix} = \begin{pmatrix} 78 & 54 & 99 \\ 104 & 72 & 132 \\ -26 & -18 & -33 \end{pmatrix}$

   (d) Since AB is 1×1 and C is 3×3 the product $(AB)C$ is not defined.

9. (a) $(AB)^T = \begin{pmatrix} 7 & 10 \\ 38 & 75 \end{pmatrix}^T = \begin{pmatrix} 7 & 38 \\ 10 & 75 \end{pmatrix}$

   (b) $B^T A^T = \begin{pmatrix} 5 & -2 \\ 10 & -5 \end{pmatrix} \begin{pmatrix} 3 & 8 \\ 4 & 1 \end{pmatrix} = \begin{pmatrix} 7 & 38 \\ 10 & 75 \end{pmatrix}$

12. $\begin{pmatrix} 6t \\ 3t^2 \\ -3t \end{pmatrix} + \begin{pmatrix} -t+1 \\ -t^2+t \\ 3t-3 \end{pmatrix} - \begin{pmatrix} 6t \\ 8 \\ -10t \end{pmatrix} = \begin{pmatrix} -t+1 \\ 2t^2+t-8 \\ 10t-3 \end{pmatrix}$

Appendix I

15. Since $\det \mathbf{A} = 0$, $\mathbf{A}$ is singular.

18. Since $\det \mathbf{A} = -6$, $\mathbf{A}$ is nonsingular.
$$\mathbf{A}^{-1} = -\frac{1}{6}\begin{pmatrix} 2 & -10 \\ -2 & 7 \end{pmatrix}$$

21. Since $\det \mathbf{A} = -9$, $\mathbf{A}$ is nonsingular. The cofactors are
$$\begin{array}{lll} A_{11} = -2 & A_{12} = -13 & A_{13} = 8 \\ A_{21} = -2 & A_{22} = 5 & A_{23} = -1 \\ A_{31} = -1 & A_{32} = 7 & A_{33} = -5. \end{array}$$

Then
$$\mathbf{A}^{-1} = -\frac{1}{9}\begin{pmatrix} -2 & -13 & 8 \\ -2 & 5 & -1 \\ -1 & 7 & -5 \end{pmatrix}^T = -\frac{1}{9}\begin{pmatrix} -2 & -2 & -1 \\ -13 & 5 & 7 \\ 8 & -1 & -5 \end{pmatrix}.$$

24. Since $\det \mathbf{A}(t) = 2e^{2t} \neq 0$, $\mathbf{A}$ is nonsingular.
$$\mathbf{A}^{-1} = \frac{1}{2}e^{-2t}\begin{pmatrix} e^t \sin t & 2e^t \cos t \\ -e^t \cos t & 2e^t \sin t \end{pmatrix}$$

27. $\mathbf{X} = \begin{pmatrix} 2e^{2t} + 8e^{-3t} \\ -2e^{2t} + 4e^{-3t} \end{pmatrix}$ so that $\dfrac{d\mathbf{X}}{dt} = \begin{pmatrix} 4e^{2t} - 24e^{-3t} \\ -4e^{2t} - 12e^{-3t} \end{pmatrix}$.

30. (a) $\dfrac{d\mathbf{A}}{dt} = \begin{pmatrix} -2t/(t^2+1)^2 & 3 \\ 2t & 1 \end{pmatrix}$

(b) $\dfrac{d\mathbf{B}}{dt} = \begin{pmatrix} 6 & 0 \\ -1/t^2 & 4 \end{pmatrix}$

(c) $\displaystyle\int_0^1 \mathbf{A}(t)\,dt = \begin{pmatrix} \tan^{-1} t & \frac{3}{2}t^2 \\ \frac{1}{3}t^3 & \frac{1}{2}t^2 \end{pmatrix}\bigg|_{t=0}^{t=1} = \begin{pmatrix} \frac{\pi}{4} & \frac{3}{2} \\ \frac{1}{3} & \frac{1}{2} \end{pmatrix}$

(d) $\displaystyle\int_1^2 \mathbf{B}(t)\,dt = \begin{pmatrix} 3t^2 & 2t \\ \ln t & 2t^2 \end{pmatrix}\bigg|_{t=1}^{t=2} = \begin{pmatrix} 9 & 2 \\ \ln 2 & 6 \end{pmatrix}$

(e) $\mathbf{A}(t)\mathbf{B}(t) = \begin{pmatrix} 6t/(t^2+1) + 3 & 2/(t^2+1) + 12t^2 \\ 6t^3 + 1 & 2t^2 + 4t^2 \end{pmatrix}$

(f) $\dfrac{d}{dt}\mathbf{A}(t)\mathbf{B}(t) = \begin{pmatrix} (6-6t^2)/(t^2+1)^2 & -4t/(t^2+1)^2 + 24t \\ 18t^2 & 12t \end{pmatrix}$

89

Appendix I

(g) $\int_1^t \mathbf{A}(s)\mathbf{B}(s)\,ds = \begin{pmatrix} 6s/(s^2+1)+3 & 2/(s^2+1)+12s^2 \\ 6s^3+1 & 6s^2 \end{pmatrix}\Big|_{s=1}^{s=t}$

$= \begin{pmatrix} 3t+3\ln(t^2+1)-3-3\ln 2 & 4t^3+2\tan^{-1}t-4-\pi/2 \\ (3/2)t^4+t-(5/2) & 2t^3-2 \end{pmatrix}$

33. $\begin{pmatrix} 1 & -1 & -5 & | & 7 \\ 5 & 4 & -16 & | & -10 \\ 0 & 1 & 1 & | & -5 \end{pmatrix} \Longrightarrow \begin{pmatrix} 1 & -1 & -5 & | & 7 \\ 0 & 1 & 1 & | & -5 \\ 0 & 9 & 9 & | & -45 \end{pmatrix} \Longrightarrow \begin{pmatrix} 1 & 0 & -4 & | & 2 \\ 0 & 1 & 1 & | & -5 \\ 0 & 0 & 0 & | & 0 \end{pmatrix}$

Letting $z = t$ we find $y = -5 - t$, and $x = 2 + 4t$.

36. $\begin{pmatrix} 1 & 0 & 2 & | & 8 \\ 1 & 2 & -2 & | & 4 \\ 2 & 5 & -6 & | & 6 \end{pmatrix} \Longrightarrow \begin{pmatrix} 1 & 0 & 2 & | & 8 \\ 0 & 2 & -4 & | & -4 \\ 0 & 5 & -10 & | & -10 \end{pmatrix} \Longrightarrow \begin{pmatrix} 1 & 0 & 2 & | & 8 \\ 0 & 1 & -2 & | & -2 \\ 0 & 0 & 0 & | & 0 \end{pmatrix}$

Letting $z = t$ we find $y = -2 + 2t$, and $x = 8 - 2t$.

39. $\begin{pmatrix} 1 & 2 & 4 & | & 2 \\ 2 & 4 & 3 & | & 1 \\ 1 & 2 & -1 & | & 7 \end{pmatrix} \Longrightarrow \begin{pmatrix} 1 & 2 & 4 & | & 2 \\ 0 & 0 & -5 & | & -3 \\ 0 & 0 & -5 & | & 5 \end{pmatrix} \Longrightarrow \begin{pmatrix} 1 & 2 & 0 & | & -2/5 \\ 0 & 0 & 1 & | & 3/5 \\ 0 & 0 & 0 & | & 8 \end{pmatrix}$

There is no solution.

42. We solve

$$\det(\mathbf{A}-\lambda\mathbf{I}) = \begin{vmatrix} 2-\lambda & 1 \\ 2 & 1-\lambda \end{vmatrix} = \lambda(\lambda-3) = 0.$$

For $\lambda_1 = 0$ we have

$$\begin{pmatrix} 2 & 1 & | & 0 \\ 2 & 1 & | & 0 \end{pmatrix} \Longrightarrow \begin{pmatrix} 1 & 1/2 & | & 0 \\ 0 & 0 & | & 0 \end{pmatrix}$$

so that $k_1 = -\frac{1}{2}k_2$. If $k_2 = 2$ then

$$\mathbf{K}_1 = \begin{pmatrix} -1 \\ 2 \end{pmatrix}.$$

For $\lambda_2 = 3$ we have

$$\begin{pmatrix} -1 & 1 & | & 0 \\ 2 & -2 & | & 0 \end{pmatrix} \Longrightarrow \begin{pmatrix} 1 & -1 & | & 0 \\ 0 & 0 & | & 0 \end{pmatrix}$$

so that $k_1 = k_2$. If $k_2 = 1$ then

$$\mathbf{K}_2 = \begin{pmatrix} 1 \\ 1 \end{pmatrix}.$$

45. We solve

$$\det(\mathbf{A} - \lambda \mathbf{I}) = \begin{vmatrix} 5 - \lambda & -1 & 0 \\ 0 & -5 - \lambda & 9 \\ 5 & -1 & -\lambda \end{vmatrix} = \begin{vmatrix} 4 - \lambda & -1 & 0 \\ 4 - \lambda & -5 - \lambda & 9 \\ 4 - \lambda & -1 & -\lambda \end{vmatrix} = \lambda(4 - \lambda)(\lambda + 4) = 0.$$

If $\lambda_1 = 0$ then

$$\begin{pmatrix} 5 & -1 & 0 & | & 0 \\ 0 & -5 & 9 & | & 0 \\ 5 & -1 & 0 & | & 0 \end{pmatrix} \implies \begin{pmatrix} 1 & 0 & -9/25 & | & 0 \\ 0 & 1 & -9/5 & | & 0 \\ 0 & 0 & 0 & | & 0 \end{pmatrix}$$

so that $k_1 = \frac{9}{25} k_3$ and $k_2 = \frac{9}{5} k_3$. If $k_3 = 25$ then

$$\mathbf{K}_1 = \begin{pmatrix} 9 \\ 45 \\ 25 \end{pmatrix}.$$

If $\lambda_2 = 4$ then

$$\begin{pmatrix} 1 & -1 & 0 & | & 0 \\ 0 & -9 & 9 & | & 0 \\ 5 & -1 & -4 & | & 0 \end{pmatrix} \implies \begin{pmatrix} 1 & 0 & -1 & | & 0 \\ 0 & 1 & -1 & | & 0 \\ 0 & 0 & 0 & | & 0 \end{pmatrix}$$

so that $k_1 = k_3$ and $k_2 = k_3$. If $k_3 = 1$ then

$$\mathbf{K}_2 = \begin{pmatrix} 1 \\ 1 \\ 1 \end{pmatrix}.$$

If $\lambda_3 = -4$ then

$$\begin{pmatrix} 9 & -1 & 0 & | & 0 \\ 0 & -1 & 9 & | & 0 \\ 5 & -1 & 4 & | & 0 \end{pmatrix} \implies \begin{pmatrix} 1 & 0 & -1 & | & 0 \\ 0 & 1 & -9 & | & 0 \\ 0 & 0 & 0 & | & 0 \end{pmatrix}$$

so that $k_1 = k_3$ and $k_2 = 9k_3$. If $k_3 = 1$ then

$$\mathbf{K}_3 = \begin{pmatrix} 1 \\ 9 \\ 1 \end{pmatrix}.$$

Appendix I

48. We solve

$$\det(\mathbf{A} - \lambda\mathbf{I}) = \begin{vmatrix} 1-\lambda & 6 & 0 \\ 0 & 2-\lambda & 1 \\ 0 & 1 & 2-\lambda \end{vmatrix} = \begin{vmatrix} 1-\lambda & 6 & 0 \\ 0 & 3-\lambda & 3-\lambda \\ 0 & 1 & 2-\lambda \end{vmatrix} = (3-\lambda)(1-\lambda)^2 = 0.$$

For $\lambda = 3$ we have

$$\begin{pmatrix} -2 & 6 & 0 & | & 0 \\ 0 & 0 & 0 & | & 0 \\ 0 & 1 & -1 & | & 0 \end{pmatrix} \Longrightarrow \begin{pmatrix} 1 & 0 & -3 & | & 0 \\ 0 & 1 & -1 & | & 0 \\ 0 & 0 & 0 & | & 0 \end{pmatrix}$$

so that $k_1 = 3k_3$ and $k_2 = k_3$. If $k_3 = 1$ then

$$\mathbf{K}_1 = \begin{pmatrix} 3 \\ 1 \\ 1 \end{pmatrix}.$$

For $\lambda_2 = \lambda_3 = 1$ we have

$$\begin{pmatrix} 0 & 6 & 0 & | & 0 \\ 0 & 1 & 1 & | & 0 \\ 0 & 1 & 1 & | & 0 \end{pmatrix} \Longrightarrow \begin{pmatrix} 0 & 1 & 0 & | & 0 \\ 0 & 0 & 1 & | & 0 \\ 0 & 0 & 0 & | & 0 \end{pmatrix}$$

so that $k_2 = 0$ and $k_3 = 0$. If $k_1 = 1$ then

$$\mathbf{K}_2 = \begin{pmatrix} 1 \\ 0 \\ 0 \end{pmatrix}.$$

51. Let $\mathbf{A} = \begin{pmatrix} a_{11} & a_{12} \\ a_{21} & a_{22} \end{pmatrix}.$

Then

$$\frac{d}{dt}[\mathbf{A}(t)\mathbf{X}(t)] = \frac{d}{dt}\begin{pmatrix} a_1 & a_2 \\ a_3 & a_4 \end{pmatrix}\begin{pmatrix} x_1 \\ x_2 \end{pmatrix} = \frac{d}{dt}\begin{pmatrix} a_1 x_1 + a_2 x_2 \\ a_3 x_1 + a_4 x_2 \end{pmatrix} = \begin{pmatrix} a_1 x_1' + a_1' x_1 + a_2 x_2' + a_2' x_2 \\ a_3 x_1' + a_3' x_1 + a_4 x_2' + a_4' x_2 \end{pmatrix}$$

$$= \begin{pmatrix} a_1 & a_2 \\ a_3 & a_4 \end{pmatrix}\begin{pmatrix} x_1' \\ x_2' \end{pmatrix} + \begin{pmatrix} a_1' & a_2' \\ a_3' & a_4' \end{pmatrix}\begin{pmatrix} x_1 \\ x_2 \end{pmatrix} = \mathbf{A}(t)\mathbf{X}'(t) + \mathbf{A}'(t)\mathbf{X}(t).$$

54. Since

$$(\mathbf{AB})(\mathbf{B}^{-1}\mathbf{A}^{-1}) = \mathbf{A}(\mathbf{BB}^{-1})\mathbf{A}^{-1} = \mathbf{AIA}^{-1} = \mathbf{AA}^{-1} = \mathbf{I}$$

and

$$(\mathbf{B}^{-1}\mathbf{A}^{-1})(\mathbf{AB}) = \mathbf{B}^{-1}(\mathbf{A}^{-1}\mathbf{A})\mathbf{B} = \mathbf{B}^{-1}\mathbf{IB} = \mathbf{B}^{-1}\mathbf{B} = \mathbf{I}$$

we have

$$(\mathbf{AB})^{-1} = \mathbf{B}^{-1}\mathbf{A}^{-1}.$$